Agee

pick PRESERVE serve

salter
SCHUYLER
$8/kg

CHRIS FORTUNE

pick PRESERVE serve

Enjoying local and home-grown produce year-round

David Bateman

To my wife Tina Fortune, my son Kacey Lewis Fortune and my daughter Holly Nancy Fortune, and in memory of my twin brother, Jeremy Fortune — 'and nothing else matters'.

First published in 2008 by David Bateman Ltd,
30 Tarndale Grove, Albany, Auckland, New Zealand

ISBN 978-1-86953-689-3

Book design and typesetting: Alice Bell
Front cover image: Getty Images
Photographs by Jim Tannock: pp 2, 4, 7, 8, 11, 12, 16, 17, 19, 24, 27, 34, 36, 39, 42, 44, 55, 58, 63, 71, 77, 78, 81, 85, 86, 89, 93, 96, 104, 108, 111, 113, 117, 121, 126
Printed in China by Everbest Printing Company

CONTENTS

PRESERVING: FRESH, LOCAL, HONEST

IN TODAY'S KITCHEN, the art of preserving has largely been pushed to the back of the cupboard. Deciding which culinary style to experiment with each night — Asian, Mediterranean, French, Pacific Rim, Greek or Italian — currently takes up our attention. But before the advent of modern transport, chilling and distribution centres, the focus fell firmly on preserving: gathering fresh fruits and vegetables from your own back yard or neighbourhood, the careful process of bottling, pickling or drying, and then being able to sample the flavours and aromas of that produce when it is less naturally abundant.

Today we have access to ingredients and products from all around the world. If we are prepared to pay for them, we can have fish from Canada, snow pea shoots from South Africa, French white asparagus, and a multitude of fresh and processed products in our supermarkets and vegetable stores. But what makes us think that produce from the other side of the world is tastier or more nutritious than that from our own back yard or community? Chinese garlic is four times cheaper than local garlic bulbs, but it has no taste; Australian tomatoes are supplied all year round, and they may look good but the travel time and storage often means they have no flavour or texture.

Some summer vegetables (asparagus, marrows) are only available for three months of the year; fresh peaches are really only good for two months of the year. Yet we expect to be able to find them in shops for longer, and think it's normal that they will come from far away. Do we want to purchase produce that we have no connection to, no understanding of where it has come from, if it's spray free or organic, or if it has been grown sustainably?

Supermarkets, distributors and marketers are changing forever the way consumers eat and consume fruit and vegetables. Modern cold storage, waxing and gas flushing of produce allows them to keep and supply product way beyond its traditional life span while still maintaining its 'freshness' in the eyes of consumers. Providing food to busy cosmopolitan communities with no access to gardens or vegetable plots is a multi-billion dollar business; these communities largely have very little understanding of produce's seasonal nature, or how far it has travelled from the farm gate to the dinner plate.

Preserving local produce is the key to enjoying fruits and vegetables year-round in a healthy, economical way. Traditionally, the summer harvest of tomatoes would have been preserved through bottling or drying and enjoyed at a later date. Many of us have memories of watching our mothers or grandmothers buying fruit while it's cheap and abundant and working in a hot kitchen at the height of summer, stirring large saucepans of plums for bottling — and later devouring the fruit warmed as a delicious winter dessert. The same principle applies for smoking fish and curing meats.

I was taught about seasonal crops by my grandparents, who harvested from the vegetable garden on a daily basis — runner beans in the cool of the morning, leeks in winter, new potatoes in late spring, raspberries in summer, and fossicking for mushrooms in autumn. They lived in an age when being self-sustainable was much more central to economic survival. If they did not grow or produce it themselves, they either missed out or traded with others in the community to get what they wanted. It was fresh, local and honest.

It's great to see the beginnings of a movement that allows consumers to re-engage with their food source, to understand how it has been produced or raised, and to learn about alternative food choices. Farmers' markets are now established in communities all over the country and allow the consumer to buy direct from the people who actually tended the crops. This enables us to learn how produce has been grown, revive the seasonal aspect of

food preparation, learn how to use less familiar vegetables and fruits and, most importantly, gather food that is tasty and nutritious in an enjoyable environment.

Farmers' markets play an important role in educating whole communities on how and where food comes from, from fruit and vegetables to cheeses, meats, seafood and wild game. One of the key aspects of a farmers' market is that the food sold there must come from the same region, generally an area of one or two hundred kilometres. If it can be grown in your community, you will generally find it being sold in an atmosphere where there are no million-dollar advertising campaigns or flashing lights, but just good old-fashioned service with a smile, communication and education. This is very similar to what my grandparents did when exchanging goods with neighbours and others in the community not so many years ago.

Having travelled around the world and sampled many cuisines, I've realised we are so lucky to have what we have in this country — what's in our own back yard is of high quality and tastes great, and it is nutritionally better for us. Most importantly, it can be very economical if bought when in abundance, and eaten in season or preserved to enjoy later. One of the additional benefits of preserving is the pleasure of giving excess food to friends and loved ones, a way of nurturing them and building stronger ties through small actions.

The following is a collection of my favourite preserving recipes and techniques. They are designed for the home cook and take into account that to enjoy good food, you must put time aside to prepare it, and know a little about the origins of the product. Most equipment used for preserving is already commonly found in the home pantry and kitchen. The techniques are not difficult to pick up, and you are in control of what goes into the finished product; there is no need for additives, artificial colours or preservatives, excess sugars or fats, rancid oils, or too many ingredients. You can make as little or as much as you want, and it is an inexpensive way of using leftover produce.

The first section of this book explains the basic methods of preserving: bottling, drying, freezing, pickling, salting, smoking, and using alcohol, oil and sugar. Within each method, I'll tell you about the method, what it involves, the types of food it particularly suits, what equipment is needed, and any relevant safety and storage tips. Use these methods as a springboard — everyone has their own favourite techniques, and it's fun to discover what works best for you and your kitchen.

The book's second section is divided into the four seasons, each starting with a list of the produce you can expect to see in abundance. The recipes for each season are ordered by method, so you can see at a glance which method is being used. Seasonal availability of crops will always vary slightly, as this country's food producers are continually discovering a multitude of micro-climates. Of course, the key to preserving is finding the time — but the effort you put in today will be rewarded as you enjoy the bounty later in the year.

For convenience, at the end of the book, there's a quick produce guide, telling you when each fruit or vegetable is in season and the best methods for preserving — so if you have a hankering to preserve apricots or mushrooms, this will tell you when to start planning and which methods to use. And there is also a temperature conversion table.

One of the central philosophies of my cooking style is 'the fresh, local, honest' mantra, because the more I know about the food I use, the better educated as a cook I am on how to get the best results from it. If it is in season, then it has not been in storage; if it is grown locally, it hasn't lost nutrients on a long journey to my kitchen; and if it has been harvested at a cooler time of day without the use of sprays and herbicides, then I get a better product with which to work. This ultimately results in a delicious dish that's a pleasure to eat.

So taste, talk, discuss, exchange ideas, laugh, communicate, network and socialise — but most of all, support your local food producers, buy locally made and taste the real flavour of this country.

METHODS OF PRESERVING

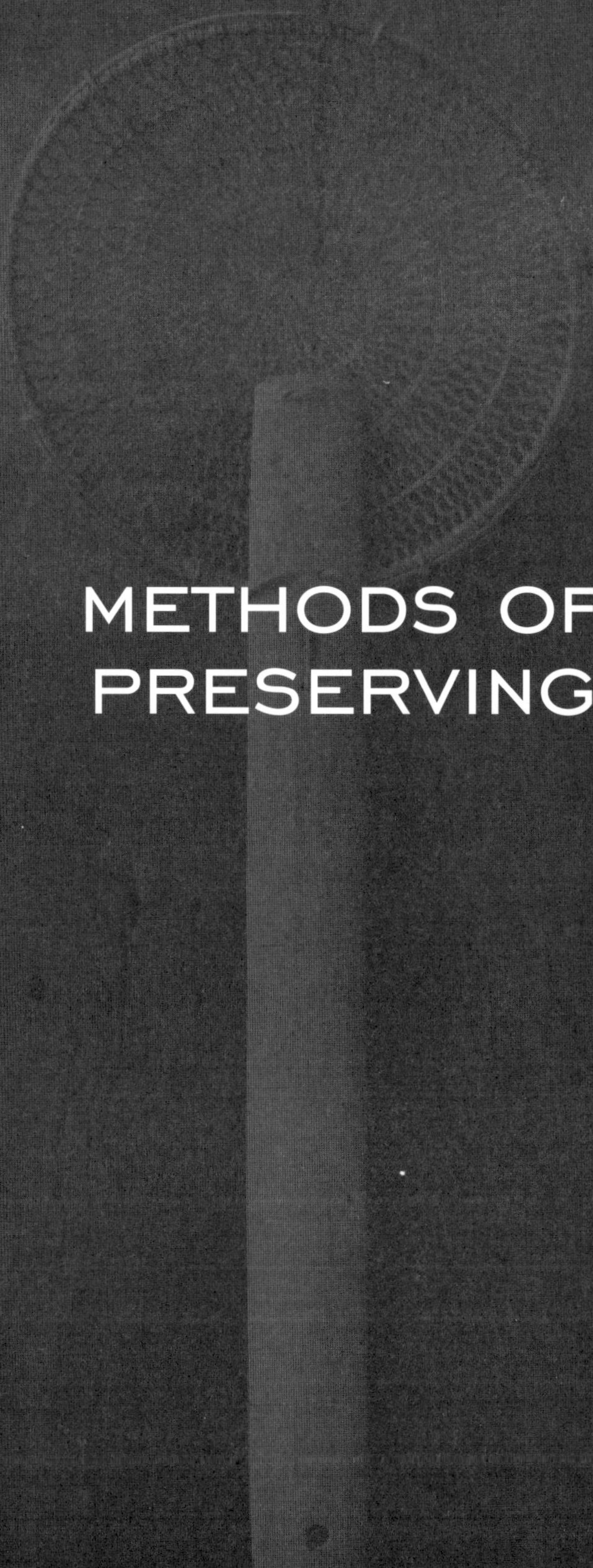

Agee

BOTTLING

Bottling is perhaps the most popular preserving method used at home today, as it suits a wide range of produce. Thanks must surely go to Napoleon, who offered a prize of 12,000 francs to whoever could devise a way to preserve food for his conquering armies — while on the march, soldiers in the late 18th century constantly battled scurvy and malnutrition. The principle is simple: the heat used in the canning process destroys the micro-organisms that cause food to spoil, and bottling then creates a vacuum in which no other bacteria can grow.

While there are many different ways to bottle edible items, and it can be more time consuming than other preserving methods, once mastered the possibilities are endless. Most vegetables, fruits and fish are suitable for bottling, especially whole or cut apples, beetroot, carrots, tomatoes and peaches. Use large-mouthed jars for large pieces of produce, and unusually shaped bottles for fancier gifts.

METHODS

RAW PACK

The raw-pack method is best used for soft fruits such as berries, plums, tomatoes, currants and cherries. Raw fruit is packed tightly into jars a third at a time, and each layer is covered with a sugar syrup solution. The syrup helps preserve the colour and texture of the fruit. Fruit juice or wine can also be used with this method. The jars, with vacuum seals and loosely tightened screw tops, are then put into a water bath — a large saucepan or preserving pan filled with enough water to come three-quarters of the way up the jars. This is then brought to the boil on top of the stove so that the fruit in the jars cooks. This will take 20–30 minutes, depending on the type of fruit used. Once the fruit is cooked and the syrup is bubbling, the jars are removed from the water bath and the screw tops screwed down as tightly as possible. A vacuum is created as the jars cool.

HOT PACK

In this method you cook the food first in order to kill all the bacteria before you seal it in the jars. It is more suited for low-acid foods such as vegetables. The ingredients are brought up to the boil or cooked in a large saucepan or preserving pan and then packed hot into sterilised jars and sealed with vacuum seals and screw tops — you do not have to process any further. A vacuum will be created as the jars cool. Peaches, pears, apples, grapes and cassoulets are most suited to this method. You must ensure that the jars are sterile and you follow the strictest hygiene standards to avoid any outside contamination.

OVERFLOW

The overflow method is commonly used for pickles or brines as there is no heat involved in the process. You must use clean, sterilised jars. The ingredients are packed into the jars and the liquid added so that it fills to the very top and slightly overflows the jar. The rim is wiped clean and the jar sealed. This method is most suited to olives, capers, courgettes and pickled onions where a brine or vinegar, which makes it impossible for bacteria to grow, is used. The vinegar must be of the best quality. Sometimes the vegetables will suit salting to draw out any excess moisture. The seals must be vinegar-proof and finished preserves kept in a cool, dark place.

PRESSURE BOTTLING

This is another method best for produce with low levels of acid, such as orange, quince and apples. You must use suitable steam-pressure glass and seals for this method. The product is packed into sterilised jars, sealed and then cooked under pressure in a pressure cooker. You can use either the hot-pack or cold-pack method to prepare the produce first, but you will find you can fit more into the jars using the hot-pack method as the ingredients will have softened. The jars are placed into the pressure cooker, which is then sealed and held under pressure for a set time. Refer to your instruction book for specific details on the time and pressure required to ensure that you have safely processed the food.

HOW TO SEAL YOUR CONTAINERS

It's best to use proper preserving jars with fitting lids for all of your preserves especially when pressure is involved. Many second-hand jars will not seal properly under pressure and may break when exposed to excessive heat for a long period of time. Do not use any lids that are lined with cardboard as they will not make a good enough seal for long periods of storage. Preserving jars, vacuum seals and screw tops are also available from most supermarkets. To ensure that you have successfully preserved the produce, check that the centre of the seal is depressed (pulled down) and does not 'give' when pressed.

Paraffin wax is another popular way of sealing jams and chutneys. Make sure that the inside of the jar rim is clean and that any moisture that would prevent the wax from sticking has been removed. Pour the hot wax directly onto the product and leave it to set.

Paper seals are ideal if you are only preserving for a short time. Egg white or gum is used to stick greaseproof paper to the jar. Plastic film or moistened cellophane tops (available from most supermarkets), held in place with a rubber band, can also be used. Always ensure that the neck of the jar is clean so that you can get a good seal. Small jars of jams are best suited to these methods of sealing.

EQUIPMENT

There are a large variety of jars and bottles that can be used for bottling. It's possible to recycle commercial glass jars for some uses, such as pickling. But when exposing glass to the high temperatures of a water bath or pressure bottling, it's safer to use sturdy, quality bottling jars. These come in a host of different sizes, with small and large necks for ease of filling.

Make sure all your equipment is clean and jars sterile. To sterilise your glass jars correctly, they must sit at 110°C for at least five minutes. This can be done either in a water bath or by washing and placing into an oven.

You will find most of the equipment you need for bottling already in your kitchen, but here is a comprehensive list of what you may need:

Bottles, jars and lids
Clean kitchen cloths
Colander
Food mill or mouli
Food processing equipment
  — slicer, blender or food processor
Kitchen scales
Kitchen timer
Large bowls
Large saucepan (for water bath)
Large wooden spoons
  and slotted spoons
Measuring cups and spoons
Paring and chopping knives
Preserving pan
Sieve
Soft-bristled
  scrubbing brush
Tongs
Wide-neck funnel

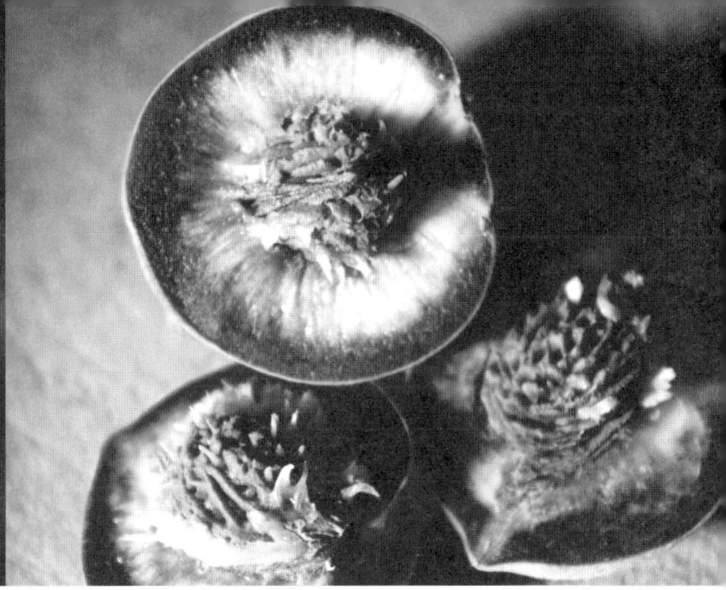

COOK'S TIPS

- Vegetables such as cucumbers, aubergines and capsicums will lose their flavour and crispness when cooked and bottled. They are more suitable for pickling. The vinegar base will provide enough acidity to stop bacteria forming as long as you keep them in a sterile environment.

- All bottled preserves are best kept in a cool, dark place to prevent light from discolouring and breaking down the contents. Shelf-life ranges from 3 months for fruit butters; 12 months for jams, jellies, pickles and conserves; 18 months for chutneys and relishes; and 2–3 years for fruits in alcohol. The better the quality of ingredients used and the stronger the preserving liquid, the longer the preserves will keep. But because all fruits and vegetables are seasonal, it is really best that they be replenished each year, so that you can enjoy them when they are at the highest level of taste and quality.

- Ensure the fruit is free of bruises and blemishes before bottling.

- If the seal or lid of a jar has 'blown', i.e. it bulges, do not take any risks — dispose of it.

- Sugar is a very important preservative, so replacing it with artificial sweeteners is not recommended. You can substitute honey or treacle instead, but be aware that it will not set as easily.

- Use only non-reactive pots and pans with heavy bases. Brass, copper, iron and aluminium tend to impart an unpleasant metallic flavour to preserves.

DRYING

Pasta, noodles, lentils, tomatoes, mushrooms, chillies, herbs, capsicums — the list is endless. If you look in your pantry, you will find countless items that have been dried. It's simple: remove the moisture and you create an atmosphere where potentially harmful organisms can no longer live. Drying is the cheapest, simplest and oldest form of preserving in culinary history — back in 12,000 BC food was being dried simply by laying it in the sun. Dried produce, both light in weight and nutritious, also sustained many early adventurers and explorers, just as it does today in the form of the fruit roll-up snacks we put in our children's school lunch boxes. And it is still used extensively as a method for preserving many crops and grains around the world. By using natural heat, wind or dehydrators, drying freshly harvested crops from the back garden or a farmers' market provides you with an abundant food supply at any time of the year.

While most produce can be dried, the best vegetables to use include mushrooms, peas, tomatoes, corn and beans. Fruits include apples, apricots, bananas, berries, cherries, figs, grapes, peaches, pears and plums.

METHODS

AIR-DRYING

This involves hanging the product out of direct sunlight to allow a combination of air and warmth to slowly carry moisture away. Keeping it away from sunlight will prevent loss of colour. You can dry grapes and tomatoes on the vine or plant to give a greatly concentrated result.

SUN-DRYING

Produce such as mushrooms and herbs will benefit more from direct sunlight. Ideally the temperature should be 38°C in direct sunlight with low humidity (below 60%). Spread the produce out on drying screens and turn the produce to ensure even drying.

DEHYDRATING

Dehydrating uses a combination of heat and air to dry produce. It gives a more consistent finish, and is easier than sun or air drying. In a dehydrator, you can preserve a wide variety of products in multiple layers. This is the best way to make fruit and vegetable leathers, as a dehydrator gives the most

consistent heat. There are many different types of dehydrators on the market: the most popular and efficient have different heat setting points, from 30°C to 65°C, which enables you to dry a wide variety of produce. The food is prepared and layered on trays and then subjected to an even flow of air over a long period of time. This is a nutritious and healthy preserving option that involves no added sugar or vinegars.

OVEN-DRYING

Depending on your oven (gas or electric), you can leave produce to dry in the oven for as little as 4 hours or as long as 24 hours. Remember to prop the door open a little to allow moisture to escape, and keep a close eye on the temperature to ensure it is not fluctuating greatly.

EQUIPMENT

Airtight plastic storage containers
Chopping boards
Dehydrator (available from good food stores)
Glass jars with sealable lids
Oven racks and non-stick paper
Sharp knives
String for hanging

COOK'S TIPS

• When drying fruits you will be removing 80% of the moisture, and for vegetables, approximately 90%. Be prepared to take the time to allow it to dry correctly, otherwise your product will go 'stale' or 'mouldy' — caused by bacteria being able to grow in a moist environment. Mould shows as white or black spores which appear on the surface of the food product. Dried food should be kept in an airtight container to extend its shelf-life and can be freshened by placing it back into the oven for several minutes if it has gone a little stale or soft.

• Dip sliced pip fruit into water and lemon juice, so it retains its colour once dried.

• Keep all your leftover cherry stalks, spread them on a tray and dry them, turning occasionally — they're great for making tea.

• Make fruit leathers with overripe fruit to use up excess product. Dried fruits and leathers are a great addition to the kids' lunch boxes, as they contain no added sugar.

• Herbs can be hung from a line in a sunny corner of the kitchen. They'll be out of the way and you can use them all year round.

• To keep flying insects away, cover drying produce with cheesecloth or another fine mesh cloth.

• Stored correctly, dried produce will last for years as long as you keep the moisture away from it, but it's best to replace it every season as new produce arrives. To maximise the shelf-life, keep in an airtight container in a cool, dark place.

FREEZING

Freezing is a very new form of preserving in the history of cooking, and it's all thanks to electricity, which has definitively changed the way we consume produce today. Although freezing has already been practised in many northern European countries for millenniums, by literally leaving produce outside in the snow, most people do not live in such conditions and rely on their deep freezers to provide these temperatures. Freezing will not kill bacteria: it merely slows down its growth, and postpones the spoilage of food products. While freezing is convenient, electricity costs make it one of the more expensive ways to preserve — however this is offset by its effectiveness in preserving the nutrients, texture, colour and flavour of produce.

Most food products are suitable for freezing if prepared appropriately. Eggs in the shell do not freeze well, as the liquid expands and will crack the shell; this can be solved by breaking the eggs into a bowl or ice cube tray and then freezing. Buttermilk, cream cheese, sour cream and whipped butter do not freeze well, as they will separate or become mushy when thawed.

METHOD

The key to the success of long-term freezing is to remove air, either by wrapping in newspaper or heavy plastic, storing in airtight containers, or layering with cardboard. Freezing works by turning liquids into a solid, so expansion will occur while freezing and extra room must be left in containers that hold liquids. A good labelling system will ensure you can easily identify each product and use it within the recommended timeframes, as freezing does not kill bacteria. Again, you should be using the freshest foods available to maintain a good result, and set your freezer at –17°C or below.

To prevent freezer burn, remove all the air pockets and wrap produce snugly. While freezing slows the growth of enzymes and bacteria in the product, it is throughout the thawing stages that you must be most careful. Thaw only in your refrigerator: this way it comes up to temperature slowly and fibre cells will not be damaged or broken, therefore retaining the moisture that had been frozen.

EQUIPMENT

While no specialised equipment is needed, thick freezer paper is recommended for wrapping.

Plastic trays and airtight containers
Zip-lock bags for smaller items
Waterproof labelling system

COOK'S TIPS

- Freeze produce in single layers, placing a sheet of plastic between each layer so you can remove small amounts at a time.

- Remove excess fat from meat and poultry, as this is the first part of meat to go rancid.

- Use uniformly sized trays and containers to maximise the freezer's interior space.

- Check your air filter every 6 months to ensure your freezer is not working harder than it needs to be.

- Fresh herbs and puréed fruits can be frozen in ice cube trays for ease of use.

- Larger cuts of meat will take 1–2 days to thaw in the refrigerator, so plan menus in advance.

- Do not refreeze high-risk product such as shellfish, seafood and poultry.

- Stone fruits are best frozen with stones removed for ease of use later. You can layer the produce on sheets of plastic film so that you can remove as little or as much as you like.

- Berry fruit is best frozen individually so that it does not stick together. Spread the fruit on plastic film or trays and then freeze; once frozen, you can tip the fruit into a suitable storage container.

PICKLING

Little did you know that the pickled green substance in today's Big Macs and Whopper Burgers originated some four millennia ago, when pickled cucumbers or gherkins were valued for their healing properties and as an important beauty aid. Pickling was once considered a necessity for preserving, but since the advent of refrigeration or freezing, pickling has become an acquired taste. While the Chinese pickle garlic, daikon and cucumbers, Europeans favour walnuts, eggs, lemons and olives. Pickling works because bacteria and microbes can't survive in the acetic acid that's present in vinegar — hence a long shelf-life is achieved.

Pickling is most suited to hardy vegetables and some meats (such as pork). Pickled onions, cucumbers, shallots, cabbage, ginger, melons and mushrooms are just a few. Pickling is a very simple way to preserve the texture of garden produce so that it can be enjoyed right throughout the year.

METHODS

DIRECT PICKLING

The food is covered in vinegar of more than 5% pure acetic acid. The flavours are sharp and fresh, and may or may not contain sugar.

INDIRECT PICKLING

Food is salted at a level that keeps harmful micro-organisms at bay but allows the lactic acid in bacteria to build up. This is the process for crock-cured pickles and sauerkraut.

In both methods, after cleaning the vegetables or fruits, layer or pack them (be careful not to over-pack) in a sterile vessel. Ensure all your equipment is clean and jars sterile. To sterilise your glass jars correctly, they must sit at 110°C for at least 5 minutes. This can be done either in a water bath or by washing and placing in a hot oven. Some vegetables, such as beans and beetroot, are better blanched prior to pickling to ensure that the vinegar penetrates into the product. Cover with vinegar and any herbs and spices, seal and store in a cool, dark place. Refer to the bottling section for more information on ways to seal containers.

EQUIPMENT

Baking paper
Glass jars
Large non-metallic bowls
Tongs
Funnel
Kitchen towels
Sieves
Wooden spoons

COOK'S TIPS

• Pickled cherries or cranberries add seasonal colour to the Christmas table.

• When using fresh capsicums in your everyday cooking, keep a jar of vinegar in the pantry and add any unused tops and bottoms of red and green capsicums. This is pickling in its simplest form — and you will always have salad capsicums on hand to use later.

• Always remove pickles with tongs or a sterile utensil, not your fingers, as you will add bacteria to the pickling solution.

• Experiment with different flavoured vinegars — spiced, red wine, white wine, cider, brown. Use a vinegar that suits the type of vegetable or fruit you are pickling.

• Use even-sized produce that is clean and free from any bruising or skin damage.

• Keep pickles in a dry, cool place and place in the refrigerator once opened.

• Place baking paper or a small weight, such as a saucer, on top of the pickled product in the vinegar to ensure the product remains submerged at all times. It is exposure to bacteria that causes pickles to go sour.

• Most vegetable pickles will be ready after 2 weeks, but fruits are better when kept for 1–2 months before eating. Using pure vinegar will increase the shelf-life of the product, as opposed to thinning out the vinegar with water or oil. Pickles will last for 12–24 months and become stronger the more they age in the brine or vinegar.

Quince
Chutney
golden harvest
mason

SALTING

Take a second look at the humble salt which graces our tables, for we seldom give it the consideration it deserves. Sodium chloride has shaped civilisations, changed the history of humankind, and has been the backbone of world trade for thousands of years. Little did the British know that Gandhi's peaceful salt march in 1930 would eventually lead to the overthrow their rule in India and allow salt to be available in most dining rooms today.

Salting is one of the oldest methods of preserving meats and fish, as it was the best way to keep proteins from going rancid. Salt cod and anchovies are two of the most recognised examples of this in the world, closely followed by ham, pork and cabbage.

With the invention of refrigeration, salt took a back seat as a preserving method and is now viewed primarily as a seasoning alongside its cousin, pepper. Salt is still an effective preserving agent, but its second function of concentrating the flavour of food by removing tasteless water is what makes it a popular choice in many cuisines today.

METHODS

You can 'dry salt', which involves packing salt around and over the product to draw out moisture, or you can 'wet salt' by dissolving salt in water to create brine. Using weights to apply pressure on large cuts of meat while salting allows moisture to dissipate. When using large crocks or earthenware jars for holding the preserves, make sure that they have been cleaned and sterilised by either adding boiling water and letting stand for 5 minutes, or by heating them in the oven at 110°C for 10 minutes before using.

EQUIPMENT

Cheesecloth or muslin, to cover
Crocks or earthenware vessels, for storing
Curing salt
Weights

COOK'S TIPS

• Pure sodium chloride or canning salt is preferred — table salt contains iodine and anti-caking agents that will discolour the final product. There are over 200 different varieties of salt available on the market today. When purchasing curing salt, be aware of the different chemicals it may contain as it varies from manufacturer to manufacturer. Sodium nitrate is more commonly used for curing bacon and hams because of its ability to preserve and keep meats 'pink'.

• Sodium nitrite (not to be confused with sodium nitrate), while once popular, is now restricted in its uses due to food safety and health considerations. While nitrites and nitrates are found naturally in a lot of food products, both fresh and processed, it is the amount that we consume that raises long-term health concerns.

• Ensure you use the freshest possible product, as this will determine the quality of the end result.

• Overuse of salt does lead to higher blood pressure, but a little is essential in the human diet.

• Cut meat or slice fish open so you are salting a thinner portion — the process will be quicker and more effective.

• Wash off any excess salt under running cold water before slicing meats or fish. Salt mixed with sugar will give you a sweet and sour finish.

• Salted meats will keep for 12–18 months when stored in a cool, dark place with good air circulation. Salted fish will last for 12 months, again stored in a cool, dark place where there is little moisture. Salted vegetables are best consumed within 8–10 months as they deteriorate quickly if left longer.

SMOKING & CURING

Preserving a cut of meat (such as a ham) or fish for a long period requires a number of steps to be followed carefully: salting (brining) and drying are the first two key steps as they remove bacteria and make it impossible for further bacteria to grow. Smoking by itself will not do this, but it will give it a great flavour as well as antioxidants that help increase the life-span of the product.

For great, tasty, ready-to-eat products that have a limited shelf-life, simply use either the hot- or cold-smoking technique described below. Hot smoke cooks as well as imparts flavour (great for chicken); cold-smoking doesn't heat or cook the product but it adds flavour and improves shelf-life (bacon is a good example). Commonly, smoking is associated with smoked salmon, beef fillets and higher-value products, although it has become popular to smoke vegetables such as garlic and mushrooms. With many types of smoking materials now available, including oak chips, manuka branches and tea leaves, smoking has come of age and can be a feature in many styles of cuisine.

More tender cuts of meat, such as loins and fillets, are most suited for smoking, as they contain less sinew and fat than others and can be sliced thinly. All types of seafood will adapt to the smoking technique, and denser vegetables such as sweet potatoes and mushrooms also hold the smoke flavour very well.

METHODS OF SMOKING & CURING

HOT-SMOKING

Heat and smoke are applied at the same time to cook the product, usually in a smoker, although you can use two roasting dishes and a rack just as effectively. While smoke will preserve the outside of the product by forming a seal, it is the brining or curing that will preserve the inside. An ideal temperature of 50–80°C enables produce to gently cook while picking up the smoky taste.

COLD-SMOKING

Cold-smoking is the process of holding the product at room temperature and introducing smoke from a different room — this imparts flavour without cooking. This is the preferred method for bacon, smoked salmon and cheeses. Cold-smoking can take from several hours to several days.

CURING

Curing is often combined with smoking to ensure you inhibit the growth of any micro-organisms. You can cure with salt or sugar as they both provide an acidic environment by drawing moisture out of the product to create an atmosphere where bacteria cannot survive. There are many recipes for curing. Some use a 50:50 ratio of salt and sugar while others use more sugar than salt. Herbs and spices can be added to give extra flavours and character to the final product.

EQUIPMENT

A vessel with a sealable lid — two old roasting dishes will suffice

Foil

Methylated spirits burner or smoky fire

Wood chips or tea leaves

Food racks or hooks

Salt and sugar for curing or brining

Patience and time

COOK'S TIPS

• While smoking will help preserve the food, it will not completely do so. Food needs to undergo several other stages of preserving to achieve a longer shelf-life. Smoked food that has not been brined or cured should therefore be treated as having the same shelf-life as fresh food.

• To avoid the bitterness associated with some smoking processes, rub sugar onto the surface of the product and rinse off afterwards.

• When smoking salmon or trout, use soft brown sugar to obtain a dark rich colour.

• Soak half the wood chips or sawdust in red wine or port for one hour prior to lighting — this will impart a rich, red wine aroma.

• Any wood chip is suitable for smoking; it's a matter of what you prefer. Some will be stronger than others and you will need a smaller quantity — apple wood, cherry wood and pine give interesting smoke flavours.

• Smoked food that has not been brined or salted needs to be stored in airtight packaging to prevent contaminating other products with smoke flavours. By wrapping in paper, and then sealing it in plastic, it will last for 6–8 days in the refrigerator.

• Smoked food products that have been cured with nitrates or brines can be kept in a cool, dark, well-ventilated place for several months.

USING ALCOHOL

The idea of using alcohol to preserve food (especially fruit and herbs) stems from the Middle Ages, when alcohol was viewed as a necessity rather than a tipple. Nothing can grow in pure alcohol, being made of fermented fruits and grains, so it is the perfect environment in which other food products can be 'embalmed'.

No heat or other additives are needed, making it a pure form of preserving still used in many parts of the world today. Everything from cherries to mushrooms can be treated in this way, with favourable results and an indefinite shelf-life.

Look for flavourful produce so the flavours will infuse into the preferred liquid. Apricots, peaches, cherries, blackcurrants, hot chillies, vanilla beans and dried herbs and spices are but a few types of produce most suited to preserving in alcohol.

METHOD

Use only the freshest ingredients — they should be unblemished and of good quality. Chop larger fruits into smaller pieces to allow the alcohol to penetrate into the product. Any alcohol can be used, but you must have a minimum of 38% alcohol to ensure bacteria cannot grow.

Ensure all your equipment is clean and jars sterile. To sterilise your glass jars correctly, they must sit at 110°C for at least 5 minutes. This can be done either in a water bath or by washing and placing into an oven. Pack product into sterile jars or bottles, top with alcohol and seal, leaving in a cool dark place to allow it to infuse. Depending on the ingredients, you should let it sit for at least 48 hours and preferably longer before using, to allow the flavours and aromas to develop in the alcohol. Refer to the bottling section for information on how to seal containers.

EQUIPMENT

Chopping board
Knives
Sterilised jars or bottles
Funnel
Non-reactive pots
Wooden spoons

COOK'S TIPS

• Use whole spices to prevent the wine or spirit from becoming cloudy.

• Shaking the bottles or tipping the vessels will help to develop flavours.

• If the finished product lacks flavour, the infusing time was too short or the ingredients were too bland.

• Experiment with different wines and spirits. Sugar can be added to offset the heat that is caused by the higher percentage of alcohol used for preserving. The higher the percentage of alcohol, the higher the heat — this gives a harsh, biting flavour.

• Most products will keep indefinitely if stored in a cool, dark place, but there will be a loss of aroma and flavour over time. Spirits will keep longer than flavoured wines, which can be made with excess fruits and vegetables. Kiwifruit, raspberry and gooseberry wine are but a few of the many varieties of wine that can be made in the kitchen.

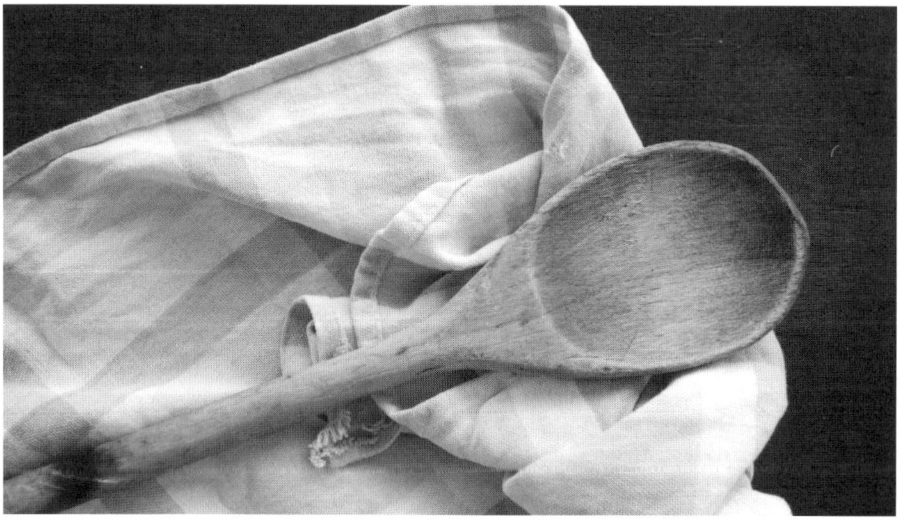

USING OIL

In vogue are the splashes of green and red oils that adorn our dinner plates at our favourite restaurants. Herbs and spices immersed in light oils and vinegars have been used for centuries as a method of preserving the seasonal crops without destroying the flavours of subtle ingredients. While preserving is achieved by submersion and the lack of oxygen, it also allows us to create eye-pleasing jars and bottles of brightly coloured essences. Immersing in oil does not impart harsh flavours as pickling does; instead you are infusing the oil with your primary ingredient, and the oil can then be used in other ways.

Herbs and vegetables with subtle flavours are most suited for preserving in oil. Rosemary, thyme, sage and parsley are but a few examples, as are olives, mushrooms, onions and capsicums.

METHOD

Gather the freshest possible ingredients. If you are using herbs, wash and dry them well. If using vegetables, you will need to salt or cook them to remove any moisture, as this is where bacteria grow. Pack the herbs or vegetables into sterile jars or crocks, seal, and leave for at least one week in a cool, dark place to mature. Refer to the bottling section for more on how to seal containers. The longer it is left, the more flavoursome the final product.

The product inside the oil, and also the oil, can then be used for different things such as cooking, drizzling on soups and food, or combined with vinegar and used as a salad dressing.

EQUIPMENT

Bottle brush
Heavy, non-reactive pot
Measuring spoons
Wax for sealing bottles
Funnel
Long-necked jars and bottles for storage
Sieve
Wooden spoons

COOK'S TIPS

• Fresh onions will turn your oil cloudy as it contains too much water; salt or dry them before preserving in oil.

• Use different flavoured oils to enhance food at the dinner table.

• If you gently warm the product before immersing it in oil, it will infuse with the oil and release more flavours.

• All products must be completely submerged in oil to ensure no bacteria can grow.

• Always use tongs or spoons to remove product from the oil, never your fingers. This will contaminate the oil.

• Excess summer herbs packed into bottles and topped with oil makes a great gift.

• Store oils in a cool, dark place: they will last for 12–18 months if stored correctly.

USING SUGAR

CONSERVES, CURDS, JAMS, MARMALADES AND SO ON

Bottled butters, jellies, preserves, jams, conserves, marmalades and others are jellied or thickened fruit products. Most are preserved by sugar. Their individual characteristics depend on the kind of fruit used and the way it is prepared, the proportions of different ingredients in the mixture, and the method of cooking.

METHODS

BUTTERS, HONEYS AND SYRUPS

Butters, honeys and syrups are preserved by sugar but not jellied. Fruit butters are sweet spreads made by cooking fruit pulp with sugar to a thick consistency. Spices are often added for extra flavour and they can be set into a mould to be turned out when serving. Apples and pears, figs, peaches and apricots all make good fruit butters.

Honeys and syrups are made by cooking fruit juice or pulp with honey to infuse its flavour into the juice to make a luxurious-tasting spread that can be spread or poured easily. Sugar is the main preserving agent as bacteria will not grow in an acidic environment. Honey, sugar and fruit are boiled together, until a light setting point is reached, then poured into sterilised bottles or jars.

CONSERVES

Conserves are jam-like products that may be made with a combination of fruits. They may also contain nuts, raisins or coconut and are not quite as thick as jams. They are much richer than conventional jams and jellies because they will have a higher fruit content. The fruit and sugar are boiled together to give a light set, and then poured into sterilised preserving jars and sealed.

CURDS

These are made with eggs, butter and sugar, and do not have a long shelf-life due to their fresh ingredients. The ingredients are combined and cooked over a hot-water bath (usually in a bowl placed over a saucepan of simmering water) so that it cooks gently and evenly so that the eggs will curdle. The bowl should not actually come into contact with the water. Pour into sterilised jars, seal and cool. They are best used within 14 days.

FRUIT CHEESES

Fruit cheeses, otherwise known as sweetmeats, contain cooked fruit pulp and equal quantities of sugar. Quince and pears make very good fruit cheese due to their high pectin content. The fruit and sugar are boiled hard until thickened and then set into containers that have straight sides so that they can be turned out easily once set. They are best left for 2 weeks to dry out and will last for 12–18 months if stored in a cool, dark place. Delicious eaten with cheese and cold, roasted meats, they make a great addition to the table.

JAMS

Jams are thick, sweet spreads made by cooking crushed or chopped fruits with sugar. Jams tend to hold their shape, but are generally less firm than jelly. This is the least labour-intensive way to process spreadable fruit. Usually the fruit is boiled hard with sugar until setting point is reached, then poured into sterilised jars and sealed. Jams can also be made by uncooked methods, using agar-agar or gelatine as the setting agent. They can be kept in the freezer or refrigerator, but need to be eaten within 2 weeks.

JELLIES

Jellies are usually made by cooking fruit juice with sugar, which is then strained by allowing it to drip through a muslin cloth into a collecting bowl beneath. Leaving the juice to drip undisturbed will produce a good, clear jelly which is firm enough to hold its shape when turned out of the container, but quivers when the container is moved. When cut, it should be tender yet retain the angle of the cut. Jelly should have a flavoursome, fresh, fruity taste and can be made with or without the addition of pectin. Boiling the fruit juice and sugar hard removes excess water. It will reach a setting temperature at approximately 105°C.

MARMALADES

Marmalades are soft fruit jellies containing small pieces of fruit or peel suspended in the transparent jelly. They often contain citrus fruit and are more like a jelly than a jam. Seville oranges, with their sharp taste and high pectin levels for setting (see page 40), are the king of marmalade fruits. Sugar and fruit is boiled hard until setting point is reached, then left to stand for 10 minutes before transferring into sterilised jars. Standing the marmalade allows the peel to distribute evenly in the jar, rather than rising to the top.

SETTING JAMS AND JELLIES

The easiest way to test for setting is to use a sugar thermometer, so you can accurately read the temperature of the product. Jams that contain sugar will set at approximately 104°C. They can also be tested by pouring a little onto a cold plate. Let it cool for several minutes and then run your finger through it; it should form a thin skin on top and leave a trail. If the jam is too thin, it will run back together. Return to the heat and boil hard, testing every few minutes until ready.

The phrase 'boil hard' is used when sugar has been added and you want to reach setting point. Bring up to the boil and instead of simmering gently, keep the heat turned up so you are removing excess moisture by evaporation.

Jams and jellies will only set if there is enough pectin in the fruit. Slightly under-ripe fruits are better, as they will contain more pectin and hold their shape and colour once preserved.

High amount of pectin	Medium amount of pectin	Low amount of pectin
Citrus fruit	Apricots	Bananas
Cooking apples (sour apples)	Blackberries	Carrots
Crab apples	Loganberries	Cherries
Cranberries	Plums	Figs
Currants	Raspberries	Grapes
Damson plums		Marrows
Gooseberries		Melons
Lemons		Nectarines
Limes		Peaches
Quinces		Rhubarb
Seville oranges		Strawberries

A simple test is to take 5 ml (1 teaspoon) of the fruit pulp or juice before you add sugar and add 15 ml (3 teaspoons) of methylated spirits. If the mixture forms a jelly clot, it has a good pectin content; if no clot is formed, you will need to add extra pectin. Lemon juice is ideal for this, as it aids the setting process and also brings out the flavour. Allow 30 ml lemon juice per 2 kg of fruit. By adding unripe apples to your preserve recipe, you can also add natural pectin — or follow the manufacturer's method when adding commercial pectins.

EQUIPMENT

As for bottling, there are a large variety of jars and bottles that can be used for holding your newly made jams and jellies and so on. It is recommended that you use proper preserving jars and bottles, although you can reuse commercial glass jars if you are not putting them into a pressure cooker. Don't forget to be very careful when pouring the very hot jams etc. into the jars, so make sure you use sturdy, good-quality jars. Those with wide necks are easier to fill. The equipment you will need is much the same as for bottling.

Bottles, jars and lids
Cellophane sealing papers and rubber bands
Clean kitchen cloths
Colander
Jelly bags
Kitchen scales
Kitchen timer
Large wooden spoons and slotted spoons
Measuring cups and spoons
Paring and chopping knives
Paraffin wax
Sieve
Soft-bristled scrubbing brush
Tongs
Wide-neck funnel

COOK'S TIPS

• Keep preserves in a cool, dark place to prevent discolouring and prolong shelf-life. Most jams, marmalades, conserves and jellies will keep for 12–18 months; butters for 14 days; curds for 2–4 months. However, they are all best eaten freshly made as the flavour does deteriorate over time.

• Best quality fruit and produce always make the best preserves.

• If the seal or lid of a jar has 'blown' or bulges, dispose of it.

• Clean equipment and sterile jars are essential. See Bottling, page 16, on sterilising jars.

• Don't replace the sugar with artificial sweeteners. You can substitute honey or treacle for some of the sugar but it may not set as easily. Sugar helps to prevent fermentation and spoilage and should be approximately 60% of the final weight of the preserve. You can replace half the sugar with honey, but it will only keep for a couple of weeks once opened.

MAKING THE MOST OF SEASONAL BOUNTY

Cooking according to the seasons, while once a necessary way of life, has largely been forgotten — it's no longer relevant to the way we eat and consume today. In a world of open-all-hours shopping and modern agriculture, we can find most items of produce all year round (although more unique products such as artichokes and asparagus, cherries and blueberries will only grow at certain times of the year).

To cook and preserve seasonally means to take the surplus of the season when it is at its abundant best, and efficiently preserve it for use later in the year. This could be as simple as freezing excess tree-ripened peaches, or bottling herbs in oil so the flavour and essence of that season is captured. The following is only a guide; adapt it to what grows in your area. Visiting your local farmers' market is a great way to learn more about seasonality and local produce.

To be eligible to sell at an authentic farmers' market, food producers must follow certain rules:

1. The product must be from a defined area, normally a one hundred to two hundred kilometre radius allowing customers to know how far it has travelled from farm gate to dinner plate.

2. The producer must be involved in the production of the product. This ensures you can get immediate answers to any questions about the best cooking methods to use, or how the product has been produced in terms of sustainability, sprays used, whether the food is organic, and so on.

3. The focus of a farmers' market is solely on selling quality food — not art and crafts or bric-a-brac.

Farmers' markets are a great way to teach new generations of children about the importance of sustainability in food production. Buying local means you eat according to the natural rhythm of the seasons, not according to the importing schedules of supermarkets or other distributors. No more cool-stored produce that looks better than it tastes.

Remember those fresh strawberries you used to eat as a child that tasted like summer, the field mushrooms that emerged in the damper autumn weather? The lemon and rhubarb puddings in the middle of winter, and the juicy cherries in spring? That is the real taste of food. To regain those experiences, one just has to preserve the memories of days gone by to be able to enjoy the art of good home cooking.

SPRING PRESERVING

AS WE EMERGE from the colder months, we look forward to the magical days of once more being surrounded by abundance. Spring is a time of fresh air and growth, so take a few simple steps to ensure you capture the joys of this harvest.

While spring may see the outdoor garden looking a little lean, modern horticulture techniques enable food producers to extend growing seasons so there is still some variety. Many winter root vegetables carry over into spring, but asparagus and early variety cherries are the real stars of this season. While many of the following fruits and vegetables are at their best in spring, some will be nearing the end of their season, whereas others, such as nectarines will only just be coming available.

SPRING FRUITS & VEGETABLES

Asparagus
Beetroot
Broad beans
Broccoli
Cabbage
Capsicums
Carrots
Cauliflower
Celery
Cherries
Chillies
Courgettes
Cucumbers
Fennel
Garlic
Globe artichokes
Leeks
Mandarins
Nectarines
Okra
Onions
Oranges
Parsnips
Peas
Rhubarb
Shallots
Spinach
Spring onions
Squash
Strawberries

SPRING RECIPES

BOTTLING

DRYING

PICKLING

SALTING

SMOKING AND CURING

USING ALCOHOL

USING OIL

USING SUGAR

CHILLI SALSA

Add a little spice to any meal with this salsa. Grilling the chillies adds a smokier flavour to the tomatoes, and the lime and coriander have a cooling effect. Wash the roots of the coriander well and chop finely, as this is where a lot of flavour resides.

20 medium-strength chillies
2 cups chopped onions
2 kg green or red tomatoes, chopped
juice of 10 limes
1 cup coriander leaves and roots
1 cup water
8 cloves garlic
1 teaspoon salt

Grill the chillies until skins blister. Allow the chillies to cool, then remove and discard the seeds and membranes. Chop roughly and add to remaining ingredients in a large saucepan. Bring to the boil and simmer for 10 minutes, ladle into hot sterilised jars and seal using the hot-pack method (see page 13).

POACHED NECTARINES

Adding crushed cardamom pods or vanilla beans will send the taste of these nectarines into another dimension.

8 cups water
600 g sugar
peel of 1 lemon (peeled with a knife)
12–18 nectarines

Bring the water, sugar and lemon peel to the boil. Halve the nectarines, remove the stones, and cut halves into quarters. Place in the sugar syrup and simmer gently for 4–6 minutes until just undercooked. Spoon the fruit into sterilised jars and cover with the syrup. Seal and store in a cool place.

SWEET & SOUR BOTTLED BEETROOT

This is a great way to preserve beetroot for the coming months. Use medium-sized beetroot and scrub them well to remove dirt. Serve the finished product with sour cream and lots of freshly ground black pepper at the dinner table, or mix with grated apples and horseradish for a fresh dip.

3 kg beetroot, well scrubbed
2 cups sugar
3 cups water
4 cups vinegar

Cook the whole beets in lots of salted water until tender — they should be soft enough to pierce with a skewer, but still offer some resistance. Allow the beets to cool, then peel and remove stems and slice. Bring the rest of the ingredients to the boil, add sliced beetroot and bring back to the boil. Bottle in sterilised jars using the overflow method (see page 14).

MARINATED GLOBE ARTICHOKES

Ensure the artichokes are at their freshest by only choosing ones that are tightly closed and firm to the touch. Once they have been pickled, you can use them in a multitude of ways: in salads, sautéed, with olives and cheese, or puréed as a dip.

300 ml white wine vinegar
6 cloves garlic, sliced
2 teaspoons salt
4 tablespoons olive oil
400 ml water
4 lemons (2 sliced, 2 juiced)
10 medium-sized globe artichokes

Bring the vinegar, garlic, salt, olive oil and water to the boil and simmer gently with 2 sliced lemons while preparing the artichokes. Remove the outer leaves of the artichokes, chop off the top third and discard. Trim the stalk to the white and then split and remove any fibrous hairs. Slice into quarters or thirds and toss with the juice of the other 2 lemons. Add to liquid and bring to the boil, simmering until tender, approximately 8–12 minutes. Pack into hot sterilised jars, and seal using the hot-pack method (see page 13).

SPICED RHUBARB

The humble rhubarb has not reached the culinary stardom of other fruits, even though it is easy to grow, but preserved rhubarb certainly stands out as a diverse and flavoursome product. It makes a delicious dessert as part of a crumble or with ice-cream.

700 ml cider vinegar
1 tablespoon salt
100 g sugar
1½ teaspoons ground cloves
1 cinnamon stick
5 thin slices of ginger
2 chillies
1 kg young rhubarb stalks, sliced

In a large saucepan, bring all ingredients except the rhubarb to the boil and simmer for 10 minutes. Add the rhubarb, bring back to the boil and simmer for 5 minutes until it just starts to go tender. Pour into sterilised jars, making sure the rhubarb is completely covered. Seal tightly.

NECTARINE FRUIT LEATHER

When making fruit leathers, overripe fruit provides a better flavour and texture. You can add and experiment with your own spices and flavours to make an endless variety of leathers. Fruits that tend to go brown when cut will need to have a little lemon juice added to them to prevent discolouration.

1 kg overripe nectarines
1 cup water
juice of 2 lemons

Peel fruit, cut into small pieces and place into water and add lemon juice. Bring to the boil, cook gently until soft and then drain well. Purée the fruit. Spread the mixture evenly on parchment paper if drying in the oven, or onto the plastic sheets of your dehydrator. The mixture should be about 3 mm thick. Try to make the centre a little thinner so that it will dry evenly. Place the fruit in your dehydrator or oven and dry — this can take anywhere from 4–10 hours, so I recommend you check it frequently, especially in the last few hours. The leathers are ready when the edges are not sticky to the touch.

DRIED BEETROOT CRISPS

A healthy, sugar-free alternative to potato chips that can be eaten as a snack or used on or in spring salads.

500 g medium to large beetroot

Peel the beetroot and with a sharp knife or mandolin, slice as thin as you can, approximately 2–3 mm. Layer onto baking paper or the racks of your dehydrator and leave to dry until crispy. The ideal temperature should be 60–80°C. Store in an airtight container.

BEETROOT RELISH

Beetroot, while bland in flavour, are rich in colour, so use this to your advantage and use them to add vibrancy when serving crusty bread or cold meats. Use rubber gloves while grating beetroot to prevent staining your hands.

1 kg grated raw beetroot
600 g grated onion
malt vinegar — enough to cover
650 g sugar
2 teaspoons salt
2 teaspoons ground black pepper
½ teaspoon nutmeg
1 tablespoon pickling spice (in a muslin or gauze bag)
2 tablespoons cornflour

In a large, heavy-based saucepan, bring all the ingredients except the cornflour to the boil and simmer for 30 minutes or until the beetroot is cooked. Remove pickling spice. Dilute cornflour in a little water and use to thicken the mixture. Pour into hot sterilised jars and seal using the hot-pack method (see page 13).

BROCCOLI & CAULIFLOWER PICKLE

After using a few common brassicas to make this tasty pickle, you will never buy commercial pickle again. Crispy and flavoursome florets will complement any sliced meats or firm cheese. This will keep for 15 months in a cool, dark place.

4 cups white wine vinegar
2 teaspoons coriander seeds
2 teaspoons cumin seeds
1 cm fresh ginger
2 heads firm broccoli
1 firm cauliflower

Bring the vinegar and spices to the boil and simmer for 10 minutes. Remove from the heat, allow to cool and then strain to remove spices. With a small knife, cut the broccoli and cauliflower into small florets and blanch in a large saucepan of salted water for 1 minute. Refresh in ice-cold water to retain colour and texture. Pack into cold sterilised jars and cover completely with the vinegar, and seal. Leave for at least 1 month before using.

LEEK & CASHEW PICKLE

Use larger leeks so you can get nice chunky slices. Clean them well to ensure there is no dirt trapped between the layers. This pickle goes very well with cold smoked chicken or brined pork, or with beef.

100 ml white wine vinegar
1 cinnamon stick
1 teaspoon allspice berries
pinch of nutmeg
2 kg large leeks
200 g cashew nuts, roasted

Bring the vinegar and spices to the boil and simmer for 12 minutes. Remove from the heat, allow to cool and then strain to remove the spices. Wash the leeks well, cut into 2–3 cm lengths, and blanch in a large pot of salted boiling water for approximately 5 minutes or until tender. Pack the leeks into sterilised jars, sprinkling roasted cashews over each layer. Cover with the cooled vinegar and seal using the hot-pack method (see page 13). Leave for at least 2 weeks before using, and refrigerate after opening.

PICKLED CUCUMBERS

4 large cucumbers, unpeeled
150 g salt
6 cups water
4 cups white wine vinegar
1 tablespoon salt, extra
2 teaspoons coriander seeds
2 teaspoons black peppercorns
5 tablespoons fresh chopped dill or fennel

Wash the cucumbers and slice into a large bowl. Dissolve the salt in the water and pour over the cucumbers. Leave for 24 hours. Bring the vinegar to a simmer, add the salt, coriander seeds and black peppercorns, and allow to cool. Drain and rinse the cucumbers, then pack in sterilised jars, sprinkling the dill or fennel between each layer. Pour in the vinegar and seal using the overflow method (see page 14).

PICKLED CHERRIES

A simple way of capturing the essence of spring in a jar, these cherries can be eaten straight from the jar or added to the Christmas ham or a red-wine gravy to give additional flavour. Once the jar is open, ensure the cherries remain below the vinegar level.

600 g sugar
10 whole cloves
8 bay leaves
900 ml white wine vinegar
1 kg firm cherries, stalks on

In a large saucepan, bring the sugar, cloves, bay leaves and vinegar to the boil and gently simmer for 10 minutes to infuse the vinegar. Leave to cool completely. Discard any cherries that are bruised or split. Pack into sterilised jars, cover with the cold spiced vinegar, and seal using the overflow method (see page 14). Leave for 10 days before eating.

CUCUMBER
$1 EACH

PICKLED ASPARAGUS

There is nothing like the taste of fresh asparagus when it's in season — but because it has a very short season, you need to ensure you put some aside to enjoy during the rest of the year. By blanching and refreshing the spears, you retain the bright green colour, while adding mustard seeds spices up the taste.

5 kg fresh asparagus, washed
2 onions, sliced
several pinches of mustard seeds
2 cups white vinegar
2½ cups water
2 teaspoons mustard seeds, extra
2 tablespoons salt
1 tablespoon pickling spice (in a muslin or gauze bag)

Blanch asparagus for 2 minutes, then drain and refresh under running cold water. Place an onion slice and a pinch of mustard seeds into the bottom of each sterilised jar. Pack asparagus spears upright into each jar. Combine vinegar, water, mustard seeds, salt and pickling spice in a saucepan and boil for 15 minutes. Remove bag and discard. Pour brine over asparagus and seal jars using raw-pack method by placing the jars in a large saucepan of water, bringing to the boil and simmering for 1 hour. Leave to cool to create a vacuum seal. Store in a cool, dark place. This asparagus can be kept for up to a year.

PICKLED SHALLOTS & GARLIC WITH BAY LEAVES

Shallots have a much more refined taste than onions and will also hold their crispiness longer. Use either banana shallots (named because of their shape) or round ones, and ensure the produce remains submerged in the vinegar at all times.

1 kg shallots
3 tablespoons rock salt
1 tablespoon black peppercorns
12 bay leaves
2 teaspoons mustard seeds
1 tablespoon salt
1 cinnamon stick
12 cloves garlic, sliced
6 cups malt vinegar

Peel the shallots, place in a large bowl, sprinkle with the rock salt and leave to drain overnight, turning occasionally. Bring the remaining ingredients to the boil and simmer for 5 minutes, then remove from the heat and allow to cool. Rinse the shallots and pack into sterilised jars. Add the spiced vinegar, ensuring there is at least 1 cm of liquid above the top of the vegetables. Seal tightly and store for at least 1 month before using.

HERBED SALT

Make good use of herbs, such as tarragon, sage, basil or thyme, when they have a good flush of spring growth on them, by salting them. Stronger flavoured herbs will work better than soft herbs such as parsley, chervil or chives.

Sea salt, to cover

Wash and pat the herbs dry. Put a half-cm layer of sea salt on the bottom of a glass vessel or wide-mouthed jar, then alternately layer the herbs with thin layers of salt until the jar is full. Leave for 4 days and you will have herb-infused salt to use in your cooking or at the table.

ROSEMARY & MANUKA-SMOKED VENISON FILLET

While farmed venison is available year-round, wild venison is at its best from late winter and into the spring. Hot-smoked venison is easy to prepare and will keep for 5 days in the refrigerator. Serve hot or cold, sliced thinly with Pickled beetroot (see page 69) or Cranberry cheese (see page 114). Alternatively, use beef or spring lamb, or try smoking the meat with thyme and orange to give a refreshing flavour.

- 3 star anise
- 2 tablespoons coriander seeds
- 1 tablespoon cumin seeds
- 1 tablespoon dried juniper berries
- 2 cinnamon sticks
- 1 tablespoon fresh thyme
- 50 ml grapeseed oil
- 1 venison fillet (approximately 700–900 g)
- 2 large handfuls rosemary sprigs
- 200 g manuka chips (or wood chips)

Dry-roast all the spices in a frying pan, cool and crush with a mortar and pestle. Heat the oil in a frying pan, seal the venison and season well with salt and pepper. Roll in the crushed spices and place on a smoking rack. Put the rosemary and manuka chips in a fish smoker or on a rack in a high-sided roasting dish lined with foil. If using a roasting dish, place in a very hot oven. Smoke the venison for about 10–12 minutes. Leave to cool and then slice thinly when required.

BRANDIED CHERRIES

A little extra effort will reward you with jars of highly flavoured cherries that can be enjoyed any time of the year. They're perfect with a soft white cheese, or vanilla ice-cream in a large bowl. Be sure to use the delicious cherry syrup as well.

500 g fresh cherries, stalks attached
300 ml water
250 g sugar
1 cinnamon stick
150 ml brandy

Prick the cherries all over with a needle. Bring the water, 100 g of the sugar, and the cinnamon stick to a simmer, then add the cherries and gently poach for 2 minutes. Remove from the heat, drain (reserving the poaching liquid) and pack the cherries into small jars. Add the remaining sugar to the poaching liquid and bring to the boil until it reaches 105°C using a sugar thermometer. Add the brandy and pour the liquid over the cherries so they are completely covered. Seal and store for at least 10 days before using.

CHERRIES IN CASSIS

This is a sweeter spiced version of pickled cherries, preserved in a sweet liqueur, that can be enjoyed at the end of the dinner with cream and a sponge cake. Use only cherries that are firm — over-ripe cherries will collapse in the brine.

800 ml rice vinegar
600 g sugar
6 cloves
100 ml cassis
1.5 kg cherries, washed, stems removed

Bring all the ingredients except the cherries to the boil. Pack cherries into hot sterilised jars, cover completely with the liquid and seal tightly. Leave for 2 months before using.

ROCKET OIL

This fresh-flavoured oil has multiple uses in the kitchen.When making flavoured oils, do not use olive oil as a base — it's best to use an oil that has no other flavours or aromas.

1 cup grapeseed oil
100 g rocket leaves

Heat the oil gently to approximately 90°C and add the rocket; the rocket will wilt in the oil. Place in a sterilised bottle. This will keep for 4–6 weeks in a cool, dark place. After a month it will start to lose the colour and flavour as it breaks down naturally.

ROSEMARY OIL

Fragrant rosemary grows well under most conditions and the new spring growth makes it particularly good for making oil. Use the oil as a marinade for meats, or dress hot, roasted potatoes with coarse sea salt and rosemary oil.

12 sprigs fresh rosemary, washed and dried
2 cloves garlic
1 teaspoon salt
2 cups grapeseed oil

In tall bottles, pack the washed and dried rosemary, then add the garlic and salt. Bring the grapeseed oil up to 40°C (approximately bath-water temperature) and pour into the bottles. Seal and leave for at least 5 days before using.

CHERRY JAM

Prepare this jam with fresh or frozen cherries. By using the excess or damaged cherries from the early season you will be able to enjoy them at any time of the year.

2 kg cherries (fresh or frozen)
juice of 3 lemons
1.5 kg sugar
¼ cup liqueur, e.g. kirsch, Cointreau, (optional)

De-stone the cherries and split the stones open. Remove the soft kernels and place them into a muslin bag. Bring to the boil with lemon juice and cherry kernels and simmer very gently for 30 minutes until the cherries are soft. Stir regularly to avoid sticking to the bottom of the saucepan. Stir in the sugar and then bring to the boil; boil rapidly for about 30 minutes then test for a set. As cherries are low in pectin they will often give a soft set. (See page 40 for information on setting jams and jellies.) Remove from the heat and stir in the optional kirsch or Cointreau if using. Ladle into clean hot jars, then seal.

ORANGE CURD

Whether spreading on toast or folding into whipped cream to be used for a cake filling, this is a simply delicious way to use up excess oranges.

2 oranges
200 g butter
400 g caster sugar
4 eggs
4 egg yolks

Finely grate the orange and then squeeze and retain the juice. Put the zest and juice into a stainless steel bowl, add the butter and sugar and place over a hot water bath, whisking until the sugar has dissolved and the butter has melted. In another bowl, lightly whisk the eggs and egg yolks, add the warmed sugar and juice and return to the water bath. Stir over the heat until the mixture thickens and lightly coats the back of the spoon. Pour into small sterilised jars and seal. Store in a cool dark place for up to 2 months or in the refrigerator once opened for 14 days.

SUMMER PRESERVING

IN SUMMER, the gardens and markets abound with fresh, tasty fruits and vegetables of the season, abundant and piled high, in assorted colours, shapes and sizes. This is the time when you can purchase produce at a good price and stock up your shelves with a colourful array of jars and bottles. There is nothing more satisfying than opening a jar of homemade preserves in the middle of winter and revisiting that summer's day when you carefully packed the hot fruits or vegetables into the jar and sealed the lid.

From strawberries to blueberries, apples to pears, watermelons to tomatoes, there is no shortage of bounty from which to choose; selecting the right time to do your preserving is the most important thing. Set aside some uninterrupted time so you can wholeheartedly focus on what you're creating — make the process of preserving an enjoyable experience rather than a chore.

Summer fruits and vegetables are quick to mature, so ensure you choose firm produce for bottling and riper produce for sauces and jam. Many of the following fruits and vegetables are at their best through the summer months, but those which have lingered over from spring will still be of good quality and suitable for preserving.

SUMMER FRUITS & VEGETABLES

Apples
Apricots
Asparagus
Aubergines
Bananas
Beetroot
Blackcurrants
Blueberries
Boysenberries
Broad beans
Broccoli
Capsicums
Carrots
Celery
Cherries
Chillies
Courgettes
Cucumbers
Fennel
Figs
Garlic
Globe artichokes
Gooseberries
Grapes
Green beans
Leeks
Limes
Loganberries
Melons
Nectarines
Okra
Onions
Oranges
Peaches
Pears
Peas
Plums
Pumpkins
Radishes
Raspberries
Redcurrants
Shallots
Silverbeet
Spinach
Spring onions
Squash
Strawberries
Sweetcorn
Tomatoes
Walnuts
Watermelon

SUMMER RECIPES

BOTTLING

DRYING

PICKLING

SALTING

SMOKING & CURING

USING ALCOHOL

USING OIL

USING SUGAR

HONEY-POACHED PEARS WITH VANILLA BEAN

These jars need to be hidden away at the back of the pantry for at least 2 months to give the pears time to macerate in the honey and vanilla. Once the flavour has infused, you have a multitude of choices when serving, including with freshly whipped cream, lemon curd or creamy ice-cream. This is also a delicious way to preserve peaches.

2 kg firm ripe pears, peeled, chopped into eighths
500 g caster sugar
200 ml honey
juice and zest of 3 lemons
1 vanilla bean, split
600 ml water

Bring all ingredients to a simmer and cook until the pears are tender. Remove the pears, carefully place them into hot sterilised jars, and bring the syrup back to the boil. Remove from heat and strain. Pour syrup over the pears, ensuring the pears are well covered, and seal the jars using the hot-pack method (see page 13).

HONEY-SPICED APRICOTS

Serve these with dollops of yoghurt for breakfast or dinner, or add a crumble topping and bake in the oven for a quick and easy dessert. If all else fails, just eat them straight from the jar.

2 kg whole apricots
cinnamon sticks and cloves for each jar
4 cups white wine vinegar
500 g honey

With a fork, prick the apricots all over and place them into cold sterilised jars. Place 1 cinnamon stick and 4 cloves in each jar. Bring the vinegar and honey to the boil in a saucepan for 5 minutes until it just starts to thicken, then pour over the apricots. Leave to cool before sealing the jars. To achieve the best flavour, leave for 1 month and use within 12 months.

SPICED BLUEBERRIES

Who can resist blueberries? These will add colour and flavour to any food at the table, especially wild meats and venison, cheeses and mixed platters.

2 cinnamon sticks
1 teaspoon whole cloves
1 teaspoon allspice berries
1½ cups red wine vinegar
8 cups blueberries
2 cups sugar

Place spices in a small piece of muslin or cheesecloth and tie securely. Bring the vinegar and spices to a simmer for 5 minutes, and then add the blueberries and carefully bring back to a simmer. Leave to cool. After 8 hours, strain the poached blueberries, retaining the vinegar, and carefully transfer berries to sterilised jars. Bring the vinegar back to the boil, add the sugar and simmer for 6 minutes until thickened. Cover the berries with the syrup and seal using the hot-pack method (see page 13).

SEMI-DRIED TOMATOES

These will be a delicious addition to the dinner table as you can keep them in a jar in the back of the refrigerator in olive or vegetable oil for that quick, easy-to-prepare meal. Covered in oil they will last for months. Remember that the oil will soak up the flavours of the tomatoes making it ideal to use for vinaigrettes and dressing.

2 kg vine-ripened tomatoes
1 teaspoon salt
1 teaspoon caster sugar
½ teaspoon freshly ground black pepper
3 tablespoons extra virgin olive oil

Line a baking tray with greaseproof paper and place halved or quartered tomatoes onto it. Season with salt, sugar and pepper and brush with olive oil. Place into a warm oven to dry for 4–12 hours depending on the size of the tomatoes. Once excess moisture has been removed they will be approximately half of their original size. They will keep for 10 days in the refrigerator, or you can put them in a jar and cover with oil and they will last for 2 months.

DRIED HERBS

To enjoy herbs all year round, spread them out on a tray in a single layer and leave in a warm place in the kitchen. Turn the herbs once a day until they are dry. Hanging them from string and suspending them to allow a good airflow also works well, and ensures the kitchen has a pleasant aroma wafting through it. Herbs that dry well include bay leaves, thyme, rosemary, parsley and oregano. If you're feeling adventurous, try other flavourful shrubs and herbs such as lemon verbena, cherry stalk, orange leaves, and chamomile flowers and blossoms.

NECTARINE & PEACH CHUTNEY

Using two of summer's best stone fruits, you'll be rewarded with a flavoursome and aromatic chutney to serve with both white and red meats. It's also great with roasted pork and root vegetables, or just by itself with cheese.

750 g peaches
750 g nectarines
2 medium onions, chopped
1 green capsicum, chopped
500 g brown sugar
2 cups white wine vinegar
3 tablespoons grated fresh ginger
1 clove garlic
2½ teaspoons cinnamon
2 teaspoons ground cloves
1 teaspoon salt
½ teaspoon black pepper

Wash and chop the fruit, and set aside with the onions and capsicum. In a large saucepan, add all the remaining ingredients and bring to a simmer, stirring frequently. Add fruit, onion and capsicum, bring to the boil, reduce heat and cook for approximately 1 hour or until thick, stirring frequently. Pour into hot sterilised jars and seal using the hot-pack method (see page 13). Leave for 2 weeks before using.

PICKLED BEETROOT

Pickled beetroot is a must at any dinner or picnic table — it goes well with salads and cheeses, or just by itself. Traditionally served with sour cream and black pepper, this simple preserve is a real taste treat.

8 medium beetroot
400 ml white wine vinegar
1½ tablespoons mustard powder
½ teaspoon salt
1 cup sugar
2 medium onions, finely chopped
1 teaspoon celery seeds
1 teaspoon dill seeds

Cover the beetroot with water and simmer until tender. Drain and reserve 300 ml of the cooking water. When the beetroot have cooled, trim their tops and bottoms and peel off the skins. Cut into 1 cm slices and place in sterilised jars. Bring vinegar and reserved liquid to a simmer with the remaining ingredients, then pour over sliced beets. Store for 2 days before using.

PICKLED PLUMS

Straight from the jar, these plums will be a culinary delight in months to come. They go well with cured and smoked products, as they provide a fruity, fresh aftertaste.

500 g sugar
zest of 1 lemon
2 whole cloves
1 knob ginger, minced
4 cloves garlic, sliced
300 ml white wine vinegar
1 kg small firm plums

Bring sugar, zest, cloves, ginger, garlic and vinegar to a simmer and then leave to cool. With a fork, prick the plums all over and add to the spiced vinegar. Simmer for 2 minutes, pack into hot sterilised jars and seal using the hot-pack method (see page 13).

PICKLED WALNUTS

This is my favourite preserve, to be used at any time of the year. It does take a few days to complete each step of the method, but the result is well worth the extra time and effort. Serve with mature cheese, or crush with a fork and mix into mayonnaise to make a tasty dressing.

1 kg green walnuts, shelled (must be no bigger than a small egg and have a soft centre)
300 g salt
water for brining
1.2 litres white vinegar
1 teaspoon allspice berries
2 teaspoons black peppercorns
3 bay leaves

With a fork, prick the walnuts all over. Dissolve 100 g of the salt in 1 litre of water, add the walnuts and leave to soak for 4 days in the brine. Drain and repeat the process, this time soaking for 3 days, then drain and repeat the soaking process for another 3 days. Drain for a final time, rinse the walnuts, spread them out on large trays and allow them to blacken for up to 48 hours before packing into sterilised jars. Bring the white vinegar, berries, peppercorns and bay leaves to a simmer and pour into the jars. Seal using the overflow method (see page 14) and leave for 4 weeks before using.

RASPBERRY VINEGAR

The tangy sweetness of summer raspberries goes ideally with thinly sliced red cabbage and cheese. Mix this vinegar with a little olive oil and you have a flavoursome seasonal vinaigrette, or serve with cured and smoked meats on an antipasto platter. This vinegar lasts for 8–12 months if kept in a cool, dark place.

500 g raspberries
2 cups red wine vinegar
1½ cups sugar

Crush the berries and leave to infuse into the vinegar for 2 days. Strain, bring to the boil and then reduce to a simmer. Add sugar, gently simmer for 5 minutes and then pour into hot sterilised bottles or jars. Seal using the overflow method (see page 14).

$12
$20

STRAWBERRY VINEGAR

This is the taste of summer in a bottle. Use it as a dressing for spinach and pear salads, or for a lighter summery dressing instead of white wine vinegar. While hydroponic strawberries will suffice, strawberries grown in the soil will have a better flavour. This recipe's simplicity makes it ideal for adapting to whatever quantity of vinegar you wish to produce.

1 part strawberries, unhulled
1 part white vinegar
¼ part white sugar

There is no need to hull the strawberries — just fill a preserving jar with strawberries and cover with the vinegar. Seal and leave for 4–5 days to mature. Strain and bring the liquid to a simmer. Add the sugar, simmer for 5 minutes and pour into sterilised jars or bottles. You can drop several pieces of whole fruit as you are sealing it. Seal using the overflow method (see page 14).

WATERMELON RIND SALAD

Serve this refreshing salad with small goods (bacon, sausages, salami), corned beef or smoked ham, as it balances the saltiness of these meats. It will get better and develop more flavours the longer that you leave it.

1 kg watermelon rind, with small amount of pink still attached
¼ cup salt
4 cups water
750 g sugar
2 cups white wine vinegar
2 cups water
1 teaspoon whole cloves
1 cinnamon stick
1 lemon or lime, thinly sliced

Slice the rind into 2-cm cubes, or 5-cm x 2-cm strips. Add the salt to the water, and leave rind to soak in the brine overnight. Drain, cover with cold water and simmer for 30 minutes or until tender. Simmer the rest of the ingredients in another saucepan for 10 minutes, then add the rind and boil until it turns translucent. Place in sterilised jars, seal using the hot-pack method (see page 13) and leave for at least 3 weeks before using.

DRY-SALTED LIMES

An easy way to use excess crops so that you can enjoy limes at any time of the year. Traditional Middle Eastern and Asian fare — a quarter lime cooked with rice or couscous adds a whole new dimension (see recipe on page 122).

20 thin-skinned limes
1 kg rock salt

Wash the limes well and make 4 slits from the top to the bottom, leaving each end intact. Open one slit on each lime and poke in a tablespoon of rock salt. Place into a sterile jar and cover with a lid. Leave in a warm location for 2–3 weeks to allow the salt to extract juice and cure the limes. When using, remove the flesh and just use the lime peel. Will keep for 12–18 months in a cool, dark place.

TEA-SMOKED CHICKEN

Cooking outside is a real summer thing to do and this recipe is an easy way to make a flavoursome summer meal using the barbecue as your smoker. If you're preparing this inside, make sure you use an extractor fan or are in a well-ventilated kitchen. Try using different tea leaves for a variety of smoke flavours, such as earl grey, green tea, rose hip and others.

4 x 200 g chicken breasts or thighs
4 teaspoons olive oil
10 tea bags, leaves reserved, bags discarded
100 g long grain rice
100 g soft dark brown sugar

Season the chicken to taste and brush with olive oil. Place the chicken on a wire rack and then onto the barbecue or into a roasting dish (you'll need another roasting dish for covering). Combine the tea leaves, rice and sugar and spread onto foil, place under the chicken. Apply heat under the foil until the tea leaves start to smoke and then cover with the barbecue lid or a second roasting dish. Leave heat on high for one minute and then turn the heat right down for 10–15 minutes until the chicken is cooked. Allow to rest for 10 minutes. Slice and serve with a fresh salad and Blackcurrant or Redcurrant jelly (see page 79) or Tamarillo chutney (see page 84) on the side.

PLUMS PRESERVED IN RUM

Eat the plums plain or with thickened cream. Make the most of the rum base as a pick-me-up sauce. These will keep for years if you hide them away; otherwise, enjoy them before other people polish them off!

230 g sugar
130 ml boiling water
1 kg firm plums
350 ml white rum (or any other alcoholic spirit)

Dissolve the sugar in the water. Prick the plums all over with a fork and simmer for 2 minutes in the syrup. Pour the plums and the syrup into hot sterilised jars and top with the rum, ensuring the plums are covered. Seal using the overflow method (see page 14) and store in a cool, dark place for at least 1 month before using.

FENNEL OIL

If you have fennel growing in your garden, let it go to seed in the summer and harvest the seeds to make this highly flavoursome oil. It has a multitude of uses, from cooking fish and garnishing soups to dressing salads.

2 tablespoons fennel seeds
400 ml grapeseed oil

Toast the seeds until golden brown under a hot grill, grind in a mortar and pestle, and add to grapeseed oil that has been warmed in a saucepan. Pour into hot sterilised jars and leave for 2 days before using. You can strain the seeds off once the flavour has infused.

BASIL OIL

A tasty addition to your salads, this oil is a great way of using up excess basil during the summer months. Mix it with tomatoes and feta cheese to create a simple, light salad. Use grapeseed oil as it has no preservatives or additives, and a fresh, healthy taste and smell.

1 cup grapeseed oil
120 g basil leaves

Heat the oil gently to approximately 90°C and add the basil. Leave to infuse for 20 minutes, then blend in a food processor until finely chopped, strain through a fine sieve and pour into a sterilised bottle. This will keep for 4–6 weeks in a cool, dark place. After a month it will start to lose the colour and flavour as it starts to naturally break down.

GOOSEBERRY & ONION MARMALADE

The sweet and sour flavours of the marmalade make it ideal for spreading thickly on toasted bread or serving with grilled or smoked chicken dishes. It can also be used as a side dish or tossed through roasted vegetables. You can use freshly picked gooseberries, or choose gooseberries that have been frozen at the height of the season.

500 g red onions, sliced
100 ml white wine vinegar
150 g raw sugar
150 ml red wine
2 cups gooseberries
1 teaspoon honey

Combine red onions, vinegar, sugar and red wine and simmer over a low heat until the mixture starts to caramelise (approximately 20 minutes). Add fresh (or frozen) gooseberries and the honey and simmer for a further 3 minutes. Spoon into hot sterilised jars and seal using the hot-pack method (see page 13).

GOOSEBERRY JAM

By adding fresh elderflowers to this simple jam, your kitchen will take on the wonderful aroma of summer. Gooseberries are a delight to some and sour to others, but they are a great addition to the breakfast table when made into jam and spread on thick slices of toast.

2 kg gooseberries, tops and tails removed
600 ml water
6 large fresh or dried elderflowers, optional
2.3 kg sugar

Place the gooseberries and the water in a large, heavy-based saucepan and bring to the boil. Simmer for 20–30 minutes or until the fruit is soft. If using elderflowers add them now. Wrap them in muslin and tie with string so that it can be removed later. The gooseberry mixture should have reduced by one-third in volume by this stage. Add the sugar and bring gently back to the boil. Boil hard for 8–10 minutes or until it reaches setting point. To test for setting point, pour a little jam onto a cold plate, let it cool, then run your finger through it — there will be a trail left through the jam if it is ready. Remove from the heat, pour into warm sterilised jars and seal using the hot-pack method (see page 13).

PLUM & WALNUT CONSERVE

Use different varieties of plums to make this thick and tasty jam — it can be used as a pie filling, or spooned into pastry to make a tart at any time of the year. You can use hazelnuts or pecans for variety.

1 kg red plums
1 kg yellow plums
2 cups water
1.5 kg sugar
300 g walnuts, dry-roasted

Bring the plums and water to the boil and simmer for 20–30 minutes, until the plums are soft and the liquid is reduced by approximately one-third. Add the sugar, bring back to the boil and let it boil rapidly until it reaches setting point (10–15 minutes). To test for setting point, pour a little conserve onto a cold plate, let it cool, then run your finger through it — there will be a trail left through the conserve if it is ready. Remove from the heat and stir in the walnuts. Pour into warm sterilised jars and seal using the hot-pack method (see page 13).

PEAR & CRYSTALLISED GINGER JAM

Pear and ginger are great partners. Crystallised ginger imparts a zing and flavour that will make this jam a family favourite.

- 2.5 kg fresh pears, peeled, cored, diced into 1 cm cubes
- 225 g crystallised ginger, roughly chopped
- 2 kg sugar
- zest and juice of 2 lemons
- 1 cup water

Bring all the ingredients to the boil in a large pot. Simmer for approximately 2 hours or until the jam sets. To test for setting point, pour a little jam onto a cold plate, let it cool for several minutes, then run your finger through it — there will be a trail left through the jam if it is ready. Pour into sterilised jars and seal using the hot-pack method (see page 13).

MINT JELLY

Homemade mint jelly earns its place at the dinner table alongside succulent roast lamb; you can also finish off your gravy with a tablespoon of jelly to give it a lift. There are many different varieties of mint now available, so ensure you only use the freshest.

3 kg cooking apples (Granny Smith, Gravenstein or
 Docherty, if you can get them)
20 sprigs mint
juice of 3 lemons
sugar to measure

Chop the apples roughly with half the mint and place in a heavy-based saucepan. Add lemon juice, and water to just cover, and simmer for 20 minutes or until the apples are soft. Spoon the apple pulp into a muslin cloth suspended over a large bowl, and leave to strain for at least 12 hours. Do not squeeze. Discard the pulp.

Weigh the liquid, and add 450 g of sugar for each 600 ml of liquid. Heat gently, stirring, until the sugar has dissolved. Boil hard for 10 to 15 minutes until the jelly reaches setting point. To test for setting point, pour a little jelly onto a cold plate, let it cool, then run your finger through it — there will be a trail left through the jelly if it is ready. Add the remaining half of the mint and pour into hot sterilised jars. Seal using the hot-pack method (see page 13).

BLACKCURRANT JELLY

This is a great breakfast treat, and also works with pork and venison dishes — a tablespoon in gravy will lift any meal to new culinary heights. The same recipe is ideal for redcurrants too.

1 kg blackcurrants (or redcurrants), stems on
1 kg sugar
1 cup of water

Combine the currants with the sugar, bring to the boil and boil hard for 10 minutes. Skim any fruit scum that comes to the surface and then strain. Working quickly, pour jelly into hot sterilised jars and seal using the hot-pack method (see page 13).

RASPBERRY JAM

Hot buttered toast or scones — no need for anything else!

1 kg fresh or frozen raspberries
1 cup redcurrant juice
1 kg sugar

Bring the raspberries and juice to a simmer, add the sugar, boil hard for 5 minutes until it reaches setting point. To test for setting point, pour a little jam onto a cold plate, let it cool, then run your finger through it — there will be a trail left through the jam if it is ready. Pour into hot sterilised jars and seal using the hot-pack method (see page 13).

AUTUMN PRESERVING

AT THIS TIME of year, the word plentiful is on the tip of the tongue. The end-of-summer crops are still in abundance, and then along comes some of the most flavoursome ingredients that nature can offer. After a long growing period, excess autumn produce is ideal for preserving, offering more taste, texture and flavour than produce from summer. Tomatoes that have failed to ripen during the hotter days are now in abundance, allowing green tomato chutneys and preserves to add another dimension to the pantry.

Time is the ingredient missing from most recipes — make time to enjoy the cooking process, don't just make it another job on your list. There's joy to be had in the sight of pots of slow-cooking green tomatoes, crocks of marinating olives, and mushrooms being pickled.

Ensure you're purchasing the best possible produce; preserving only works well when you start with good ingredients and carefully preserve them so they can be enjoyed later in the year. Many of these fruits and vegetables are also available through spring and summer, but others, especially field mushrooms, quinces, blackberries and persimmons are at their very best in autumn.

AUTUMN FRUITS & VEGETABLES

Apples
Aubergines
Avocados
Beetroot
Blackberries
Broccoli
Brussels sprouts
Cabbage
Capsicums
Cauliflower
Celery
Chillies
Courgettes
Cranberries
Cucumbers
Fennel
Figs
Gooseberries
Grapes
Green beans
Kiwifruit
Leeks
Lemons
Limes
Mandarins
Melons
Mushrooms
Okra
Olives
Onions
Oranges
Pears
Persimmons
Pomegranates
Pumpkin
Quinces
Rhubarb
Shallots
Silverbeet
Spinach
Squash
Tamarillos or tree tomatoes
Tomatoes

AUTUMN RECIPES

FIG CHUTNEY

Figs ripen very quickly, so when there is an excess freeze them in slices so you can preserve them later. Preserved ginger will give this chutney more zing and freshness when serving with roasted meats or a selection of cheeses.

1.5 kg green figs, sliced
500 g onions, sliced
100 g preserved ginger, finely chopped
2 cups malt vinegar
325 g honey
150 g raisins
¼ teaspoon cayenne pepper
1 teaspoon salt

Place all ingredients in a large saucepan and bring to a simmer, stirring so the mixture doesn't stick to the bottom. Simmer gently for approximately 2 hours until no excess liquid remains and the mixture is thick. Spoon into sterilised jars and seal using the hot-pack method (see page 13). Store in a cool, dark place.

TAMARILLO CHUTNEY

Tamarillos, or tree tomatoes, are best preserved while still firm and tart. This chutney goes well with cold-cut meats and bread.

1 kg tamarillos
500 g apples
750 g brown sugar
350 g onions, diced
120 g raisins
1 teaspoon mixed spice
1 teaspoon salt
¼ teaspoon cayenne pepper
75 g preserved ginger
2 cups white wine vinegar

Peel the tamarillos by piercing the top, covering with boiling water and leaving to sit for 2–3 minutes until the skin comes away. While warm, chop into ½-cm dice. In a large saucepan, bring all the ingredients to the boil and simmer gently for 1–1½ hours until it is thickened and no longer watery. Pour into sterilised jars and seal by using the hot-pack method (see page 13).

DRIED APPLE RINGS

Easy and quick, these are enjoyed by all ages either as a chewy snack or added to breakfast cereals. Use different varieties and colours of apples to keep the whole family happy.

- 1 kg apples
- 1 cup cold water
- 2 tablespoons lemon juice

Core and slice the apples into rings, then place in a bowl with the water and the lemon juice. Leave to stand for 15 minutes, drain well, pat dry and lay on lined baking trays. Place in a dehydrator or oven at 80–100°C and dry for approximately 4–7 hours, depending on the thickness of the slices. Cool and store in an airtight container.

KIWIFRUIT LEATHER

Easy to prepare and a great way to encourage younger people to eat their fruit in an easy roll-up. The leather will last for 8–10 months in an airtight container or plastic film in the pantry.

1 kg ripe kiwifruit
juice of one lemon
2 teaspoons honey, optional
1 cup water

Bring all ingredients to the boil and simmer until very soft, approximately 4 minutes. Spread the mixture evenly onto parchment paper if drying in the oven, or onto the plastic sheets of your dehydrator. The mixture should be about 3 mm thick. Try to make the centre a little thinner so that it will dry evenly. Place the fruit in your dehydrator or oven at approx 60–80°C. The leathers are ready when the edges are not sticky to the touch. Drying can take anywhere from 4–10 hours so I recommend that you check it frequently in the last few hours.

CINNAMON GRAPES

A sweeter version of pickled grapes, these will be a favourite at dessert time. The longer the grapes are left to macerate in the syrup, the better they will taste.

500 g brown sugar
4 cups white wine vinegar
6 cloves
6 juniper berries
1 cinnamon stick
zest of 1 lemon
1 kg grapes

Simmer all the ingredients, except the grapes, in a large saucepan for 5 minutes. Pack the washed grapes into hot sterilised jars and cover with the hot liquid. Seal using the hot-pack method (see page 13) and store for at least 2 months before using.

AUBERGINE CHUTNEY

Although there are a number of ingredients in this recipe, it's really very easy and the end result will age well. Aubergines need to be salted and drained to remove the excess moisture and bitterness. Toasting the spices and seeds will freshen them by removing moisture so the aromas and tastes are more defined.

1 kg aubergines
1 teaspoon salt
2 teaspoons turmeric

Dice the aubergine and mix with the salt and turmeric. Allow to stand for 1 hour, then fry in hot oil for approximately 4–5 minutes until golden brown. Set aside.

3 tablespoons yellow mustard seeds
1 tablespoon ground coriander seeds
2 tablespoons ground cumin seeds
1 tablespoon fennel seeds
2 chopped onions
4 cloves garlic
1 tablespoon minced ginger
½ cup white wine vinegar

Toast all the seeds until brown under a hot grill, then fry them with the onions, garlic and ginger. Remove from heat, add vinegar and add this mixture to the aubergine.

¼ cup tamarind seeds
1 cup water
3 chillies, deseeded and chopped
4 tablespoons sugar

Bring all ingredients to the boil and simmer gently for 15 minutes. Add to the aubergine mixture, pour into hot sterilised jars and seal using the hot-pack method (see page 13).

GREEN TOMATO CHUTNEY

A great recipe to use up an excess of unripened tomatoes so they can be enjoyed later in the year. Refrain from eating them for 1 month and it will allow the flavours to develop and mature.

1.5 kg green tomatoes, diced
500 g cooking apples, peeled, cored, chopped
250 g onions, peeled and chopped
125 g sultanas
250 g raw sugar
2 teaspoons salt
2 cups malt vinegar
1 tablespoon minced ginger
½ teaspoon cayenne pepper
1 teaspoon mustard powder

In a large saucepan, bring all ingredients to a simmer and gently cook for 1½ hours, stirring occasionally. When it is reduced to a thick consistency and no excess liquid remains, spoon into sterilised jars and seal using the hot-pack method (see page 13). Store in a cool, dark place for at least 1 month before using.

MUSHROOMS PRESERVED IN WHITE WINE & THYME

These mushrooms will retain all their autumn flavours while absorbing the subtle flavours of white wine and thyme. They will keep for 4 months in a cool, dark place.

4 cups white wine vinegar
2 dried red chillies
4 cloves garlic, sliced
1.5 kg button mushrooms, wiped clean with a damp cloth
2 tablespoons salt
approximately 400 ml dry white wine
fresh thyme sprigs

Bring the vinegar, chillies and garlic gently to the boil and simmer for 10 minutes. Allow to cool, then strain. Spread out the mushrooms, sprinkle with the salt and leave for 1 hour. Cook in a large saucepan for approximately 5 minutes (the salted mushrooms will make enough liquid to do this), add the spiced vinegar, bring back to the boil for 3 minutes and allow to cool. Leave overnight, then strain the liquid and weigh it. Add the same amount of white wine and bring to the boil. Place mushrooms into sterilised jars, add a sprig of thyme and cover with boiling wine and vinegar. Seal using the overflow method (see page 14) and store.

PICKLED SHALLOTS

Use large pottery crocks or jars with open mouths to store these shallots. They're a classic when paired with firm cheese and other finger foods.

1 kg shallots, peeled and trimmed
4 tablespoons coarse salt
5 cups Sweet spiced vinegar (see page 106) or any spiced vinegar available from good supermarkets

Sprinkle the shallots with the salt, leave to stand for at least 12 hours, and then drain. Pack into sterilised jars and cover with vinegar, placing a small weight on top to ensure the shallots stay submerged. Seal using the overflow method (see page 14). Leave to sit for 4 weeks before consuming.

PICKLED MUSHROOMS

Use smaller button mushrooms that are tightly closed for best results. Malt vinegar lends more flavour and colour to the final preserve. Once you open a jar of these mushrooms, make sure that they remain covered in vinegar by pressing them down into the liquid using a small sterilised plastic lid that has several holes drilled into it to allow the vinegar to flow through, placed upside down in the jar.

4 cups malt vinegar
2 teaspoons salt
½ teaspoon ground black pepper
4 shallots, finely chopped
4 sprigs fresh thyme
1.2 kg small button mushrooms, trimmed and wiped clean with a damp cloth

Gently simmer all ingredients, except the mushrooms, for 5 minutes. Spoon the mushrooms into hot sterilised jars and cover with the vinegar. Seal using the overflow method (see page 14) and store in a cool place, leaving for 1 month before devouring.

PICKLED GRAPES

An easy way to preserve the harvest from the vine; if you can find seedless varieties of grapes, all the better. By adding cloves and ginger, you will enjoy a lovely warm flavour when serving them in the middle of the winter.

300 ml white wine vinegar
50 g sugar
1 teaspoon cloves
1 teaspoon ginger
1 teaspoon salt
1 kg grapes (½ red, ½ black), firm and unblemished

Simmer all ingredients except the grapes for 5 minutes and set aside to cool. Pack the grapes into sterilised jars. Strain the pickling vinegar and pour into the jars, ensuring the grapes are covered. Seal using the overflow method (see page 14) and store in a cool, dark place.

SWEET AND SOUR TOMATOES & APPLES

These will go perfectly with white meats and garden vegetables, either as a side dish or mixed with rice or other grains to make a tasty salad.

500 g green tomatoes
4 tablespoons grapeseed oil
4 green chillies, cored and seeded
2 onions, finely diced
2 cooking apples, peeled and diced
¼ teaspoon allspice
½ teaspoon salt
4 tablespoons brown sugar
¼ cup white wine vinegar

To skin the tomatoes, pierce their skins, place in a bowl, cover with boiling water and let them stand for 3 minutes. Drain, peel while hot and chop finely. Heat the oil in a medium-sized frying pan, add the remaining ingredients, bring to the boil and simmer for 30 minutes until thick and puréed. Pour into hot sterilised jars and seal using the hot-pack method (see page 13).

CUMIN-SALTED CABBAGE

Easy to prepare, once the fermentation process has finished it will last for 3 weeks in the refrigerator and can be used hot or cold. Best served with pork and cured meats.

6 cups finely shredded cabbage
4–5 teaspoons sea salt
2 teaspoons cumin seeds

Mix all ingredients together until the cabbage starts to leak water. Place in a large, non-reactive container or glass jar, packing the cabbage in tightly. The cabbage needs to be covered with liquid, so pour in some brine (2 cups water, 1 tablespoon sea salt), if necessary, to cover. Place a weight on top to keep the cabbage submerged and allow to ferment in a cool, dry place for at least 1 week. After fermenting (bubbles have stopped), refrigerate in an airtight container.

CURED SALMON WITH FENNEL & PRESERVED LEMON

Salmon are at their prime in autumn after their growth spurt over the summer months. This recipe requires patience, as it takes several days for the salt and sugar to draw out the moisture. Once cured, you can either slice it thinly and eat it plain, or cut into larger portions and seal in a frying pan before serving.

1.5 kg salmon fillet, pin boned and skin removed
1½ cups salt
1½ cups soft brown sugar
½ cup ground black pepper
200 g fennel bulb
peel of 1 Preserved lemon, thinly sliced (see page 110)
1 tablespoon chives, chopped
1 tablespoon crème fraîche

Place the salmon fillet on a large tray and trim so it is the same thickness all the way through. This ensures it cures at an even rate. Mix salt, sugar and pepper together, spread half the mixture on top of the salmon and half underneath the salmon, and cover with plastic food wrap. Place another tray on top and put weights on top to press firmly. Leave for 24–32 hours in the refrigerator, depending on the thickness of the fillet. The thicker the fillet the longer you will have to leave it. Wash off salt mixture and slice thinly. Slice the fennel bulb and mix with the lemon skin, chives and crème fraîche to make a refreshing salad to serve with the salt-cured salmon.

WATER-CURED OLIVES

Use only freshly picked olives free of bruising or blemishes. Ninety-five per cent of olives are cured using chemicals, and you can taste the difference when you water-cure them yourself. The olive oil on top will seal the salt brine. Ensure you only remove the olives from the brine with clean spoons to avoid contamination.

5 kg black olives, well washed
sea salt
water
4 tablespoons red wine vinegar
olive oil

With a sharp knife, make a cut in the meat of the olive from top to bottom. In a large non-reactive container (plastic, not aluminium or copper) soak the olives in brine (1 part sea salt to 10 parts water). Make sure the olives are submerged by using something to weigh them down, and cover the container with a lid. Cure the olives for at least 3 weeks, stirring the olives each day and changing the brine every 4 days. Periodically, taste the olives for bitterness — they could take 5–6 weeks depending on the olives. When they taste the way you want, place the olives in jars, cover with brine (as above), add the vinegar and top with a layer of olive oil.

WINE-BARREL-SMOKED MUSHROOMS

Mushrooms are very tasty when smoked, as they absorb aromas very well. Serve with Apple & sage butter (see page 97) or Spanish onion marmalade (see page 99). Get your wine-barrel shaving from wineries or barbecue shops.

1 kg button mushrooms
2 teaspoons salt
1 teaspoon black pepper
1 tablespoon fresh thyme
2 tablespoons grapeseed oil
1½ cups wine-barrel shavings (or oak shavings soaked in 1 cup of red wine for 30 minutes)

Wash the mushrooms well and season with salt, pepper, thyme and oil. Spread out on a roasting rack and, either using a barbecue with a lid or two large roasting dishes lined with tinfoil, spread the wood chips around. The mushroom rack should be 4–6 cm above the smoking chips. Place on the heat and bring up to smoking point. Turn down the heat and leave for 10–12 minutes until mushrooms are lightly smoked and half cooked. Leave to cool and store in an airtight container or breathable bag in the refrigerator.

APPLE & BLACKBERRY CHEESE

Fruit cheeses are designed to be eaten as a spread or sliced, and go well with strong blue cheeses or when dropped into sauces and gravies at the last minute. Use only tart apples to counteract the sugar and help with the setting process.

2 kg tart green apples, peeled and cored
1 cup water
1 kg blackberries
sugar to measure

Place apples, water and blackberries in a heavy-based saucepan. Simmer for 10 minutes until soft and tender. Pass through a sieve or mouli and weigh the fruit. For every kilogram of fruit pulp, add 750 g of white sugar. Bring to the boil and boil hard for 20–30 minutes or until it starts to set. To test for setting point, pour a little cheese onto a cold plate, let it cool, then run your finger through it — there will be a trail left through the cheese if it is ready. Pour onto a greased and lined baking tray, or put into small plastic vessels, and allow to dry for 2–3 days. This will last for 8–12 months stored covered in the cupboard in a cool, dark place.

APPLE & ORANGE JELLY

By using the pips in the cooking stage, you are adding more pectin — an essential ingredient that helps jellies to set. Smaller crab apples are ideal, as they have higher pectin levels and a tartness that counteracts the sweetness of the sugar.

1 kg cooking apples, unpeeled
3 oranges
4 cups water
sugar to measure

Core the apples, reserving the pips. Zest the oranges and set aside for later use. Roughly chop the oranges, reserving the pips. Tie all apple and orange pips in a muslin bag and place in a saucepan with the apples, oranges and water. Bring to the boil and simmer for approximately 40 minutes or until soft and mushy. Discard the pips, and strain the pulp in a jelly bag or fine sieve for at least 4 hours (or overnight). Measure the juice and for every 600 ml, add 450 g sugar, bring to a simmer and boil hard for at least 10–12 minutes or until it reaches setting point. To test for setting point, pour a little jelly onto a cold plate, let it cool, then run your finger through it — there will be a trail left through the jelly if it is ready. Pour into warm sterilised jars, sprinkle the orange zest on top, and seal using the hot-pack method (see page 13).

APPLE & SAGE BUTTER

Roast pork only gets better when teamed with apple and sage butter. Alternatively, use it as a condiment or spread. The mix of sweet and savoury flavours allows you to use it in many different ways.

- 2 kg cooking apples, peeled, chopped and cored
- 1 cup sugar
- 1 teaspoon salt
- 1 teaspoon black pepper
- 2 teaspoons dried sage
- 1 medium-sized onion, diced
- 150 ml water
- 50 g butter
- 3 tablespoons white wine vinegar
- ½ teaspoon Worcestershire sauce

Mix all the ingredients together in a large, heavy-based saucepan. Bring to the boil and simmer until the apples are soft, then beat with a large whisk until smooth and creamy. Pour into warm sterilised jars and seal using the hot-pack method (see page 13). It will last for 3 months in a cool, dark place but once opened must be eaten within 7–10 days.

FIG & PORT JAM

Figs are readily available in early autumn and are ripe for picking when you see lots of bird activity around the tree. One of the more flavoursome fruits, figs contain lots of natural sugars and are best eaten on the day of picking.

- 2 kg ripe figs, stalks trimmed
- 800 g sugar
- peel of 2 lemons, thinly sliced
- 1 cup water
- ½ cup port

Roughly cut the figs and place in a large, heavy-based saucepan with the sugar, lemon peel and water. Bring to the boil and simmer until setting point has been reached. To test for setting point, pour a little jam onto a cold plate, let it cool, then run your finger through it — there will be a trail left through the jam if it is ready. Add port, boil hard for 2 minutes and ladle into hot sterilised jars. Seal using the hot-pack method (see page 13). The jam will last for 12–18 months if kept in a cool, dark place. Once opened, use within 14 days.

FIG & QUINCE CONSERVE

There is always room in the pantry for more jams and conserves, as they make ideal gifts for friends and family. People either dislike or love quince; when cooked, they take on a beautiful golden colour and will set well as they contain a high ratio of pectin.

1 kg figs, quartered
1.5 kg quinces, peeled and sliced thinly
1 kg sugar
juice and zest of 2 oranges
juice and zest of 2 lemons
2 cups water

In a large saucepan, bring all the ingredients to the boil and simmer until it reaches setting point. To test for setting point, pour a little conserve onto a cold plate, let it cool, then run your finger through it — there will be a trail left through the conserve if it is ready. Pour into sterilised jars and seal using the hot-pack method (see page 13). Keep in a cool, dark place until ready to be used.

QUINCE JELLY

Sacred to many, the quince was the symbol of happiness and fertility in Roman and Greek times. Today, it still deserves a place on the pantry shelf.

2 kg quinces, washed and roughly chopped
juice and grated zest of 3 lemons
14 cups water
sugar to measure

Bring quince, lemons and water to the boil and simmer for approximately 1 hour until tender. Place into a jelly bag and leave to strain for at least 12 hours. Discard the pulp, return the liquid to the saucepan and for every 600 ml of liquid add 500 g of sugar. Bring to a simmer and then boil rapidly for about 10–15 minutes until setting point is reached. To test for setting point, pour a little jelly onto a cold plate, let it cool, then run your finger through it — there will be a trail left through the jelly if it is ready. Pour into hot sterilised jars and seal using the hot-pack method (see page 13).

QUINCE PASTE

Traditionally this paste is thick, sliceable, and spread on fresh bread with blue cheese. Quince paste is best made with quinces that are not quite ripe as they are higher in pectin.

3 kg quinces
water
sugar to measure

Without peeling or coring, cut the quinces and place them in a large saucepan and add just enough water to cover. Simmer gently for approximately 1 hour until the fruit is soft. Pass through a sieve or mouli and measure the pulp. For every 600 ml of pulp, add 500 g of sugar, then return to the saucepan. Bring to the boil and simmer, stirring frequently until thick and a deep rose colour. To test for setting point, pour a little paste onto a cold plate, let it cool, then run your finger through it — there will be a trail left through the paste if it is ready. Pour into a shallow non-stick baking tray or plastic containers and leave to set overnight. If needed, place into a warm oven to dry over several days until firm. Quince paste will last for 12–18 months stored covered in a cool, dark place.

SPANISH ONION MARMALADE

This marmalade goes with both sweet and savoury foods. You can substitute white onion instead of red onion.

2.5 kg red onions, peeled and thinly sliced
1 cup white wine vinegar
500 g raw sugar
2 cups red wine
50 ml honey
¼ bunch sage, finely chopped

In a large, heavy-based saucepan, combine all the ingredients except for the sage and bring to the boil. Simmer for 20–30 minutes, stirring every 3–4 minutes, until the mixture starts to caramelise. Add the fresh sage, pour into hot sterilised jars and seal using the hot-pack method (see page 13).

WINTER PRESERVING

ALTHOUGH WINTER may be bleak and cold, there are fruits and vegetables that thrive at this time of the year and our eating habits have changed to reflect this. We enjoy hearty cooking, more satisfying dishes that reflect the coolness of the short days. One of the great pleasures of winter is to open a jar of home-preserved summer peaches and serve them with dollops of thickened cream. The smells and tastes of hot summer days will linger at the dinner table long after the last segment has been finished. Winter is a time for reflection and gratitude that the energy and time invested in the earlier months has paid dividends.

Citrus is at its abundant best in the cooler months; root crops will stay fresh and crisp in the ground until ready to be preserved and pickled. Cranberries and beetroot will add colour to the pantry shelves and their glowing red hues will add warmth to any meal.

Winter produce is considered to be the soul food of the year, for it must comfort and warm us more than that of other seasons. Jerusalem artichokes, chicory and rhubarb will make an interesting addition to any dinner table. Varieties of many of these fruits and vegetables are also available in other seasons, but true winter produce is best throughout the colder months.

WINTER FRUITS & VEGETABLES

Beetroot
Broccoli
Brussels sprouts
Cabbage
Carrots
Cauliflower
Celeriac
Celery
Cranberries
Fennel
Grapefruit
Jerusalem artichokes
Kale
Kiwifruit
Lemons
Limes
Mandarins
Onions
Oranges
Parsnips
Pears
Pumpkin
Rhubarb
Silverbeet
Spinach
Tamarillos or tree tomatoes

WINTER RECIPES

BOTTLING

DRYING

PICKLING

SALTING

SMOKING & CURING

USING ALCOHOL

USING OIL

USING SUGAR

GARLIC & TREACLE SAUCE

These old-fashioned flavours will go well with roasted or barbecued meats, potatoes and pumpkin. Use as a condiment at the table or as a marinade in the kitchen. This sauce lasts for 12–18 months.

10 cups brown malt vinegar
2 lemons
100 g garlic
50 g salt
2 cups treacle
500 g brown sugar
1 teaspoon powdered ginger
20 g cloves
¼ teaspoon cayenne pepper

Place all ingredients in a large saucepan and simmer for 50 minutes until thickened. Pass through a medium-sized sieve and pour into sterilised jars. Seal using the hot-pack method (see page 13).

BRAISED RED CABBAGE

Cabbages are cheap and easy to preserve. They can be eaten hot or cold, as a main vegetable dish served with poultry or game, or as a condiment. Add apple and cloves to this basic recipe if you'd like to enhance and enliven the preserve.

250 g onions
1 tablespoon crushed garlic
1 kg red cabbage, sliced thinly
150 g butter
150 g brown sugar
300 ml white wine or red vinegar
1 tablespoon salt
1 teaspoon black pepper

In a large frying pan, cook the onions and garlic until soft, add cabbage and sweat for 2 minutes. Add remaining ingredients and cook for 12 minutes until tender. Pack into sterilised jars and seal using the hot-pack method (see page 13). Leave for 2 weeks before using.

PUMPKIN LEATHER

A great snack on the run and can be kept in the pantry wrapped in plastic wrap for 8–10 months. As you dry fruits and vegetables their natural sugars condense so there is no need to add the honey if you don't want to.

1 kg pumpkin, peeled and chopped finely
2 tablespoons honey (optional, to taste)
2 cups water

Bring all ingredients to the boil and simmer until very soft, approximately 10 minutes. Pass through a sieve and spread the mixture evenly on parchment paper if drying in the oven, or onto the plastic sheets of your dehydrator. The mixture should be about 3 mm thick. Try to make the centre a little thinner so that it will dry evenly. Place the fruit in your dehydrator or oven at approx 60–80°C. The leathers are ready when the edges are not sticky to the touch. Drying can take anywhere from 4–10 hours. I recommend checking it frequently in the last few hours.

SWEET SPICED VINEGAR

Use as a base for your vinaigrettes or salad dressings, or mix with thinly sliced red cabbage and green apples to make a refreshingly simple salad. It is also equally good to use in some of the following recipes, for example, in Pickled Winter Cole pears or Pickled red cabbage (see page 108).

4 cups white wine vinegar
250 g raw sugar
1 teaspoon salt
1 teaspoon whole mixed spice
1 teaspoon peppercorns
½ teaspoon whole cloves

Bring all the ingredients to the boil and simmer for 5 minutes before pouring into warm sterilised bottles. Seal with an airtight lid and leave for at least 1 month before using.

CELERY & FENNEL RELISH

Fennel goes well with white meats and seafood, as it has a lovely aniseed flavour. Served hot or cold, this is a great addition to the pantry and will last well into spring.

8 sticks celery
2 fennel bulbs
2 onions
2 cloves garlic
8 tablespoons grapeseed oil
¼ tablespoon white wine vinegar
¼ cup sugar
juice and zest of 1 lemon

Finely chop all vegetables into a small dice. Place in a medium-sized saucepan and sweat with the oil until the onion turns translucent. Add the vinegar, sugar and lemon juice and zest, bring to a simmer, then boil hard until the liquid is reduced to a thickened syrup. Remove from the heat, pour into warm sterilised jars and seal using the hot-pack method (see page 13).

JERUSALEM ARTICHOKE & GARLIC PICKLE

Serve this treat with cheese and small goods. As an uncooked preserve, and since the preserving qualities of vinegar are diluted by the water, this is best kept in the refrigerator and consumed within 3–4 weeks. Some Jerusalem artichokes (or sunchokes) are very knobbly, so break them into small pieces to make peeling easier.

250 g artichokes, peeled and sliced thinly
1 onion, sliced
1 red capsicum, sliced
120 ml white wine vinegar
120 ml water
1 large clove garlic, sliced
½ teaspoon salt
1 tablespoon sugar
1–2 pinches cayenne pepper

Pack all ingredients into hot sterilised jars and seal using the overflow method (see page 14). Store in the refrigerator until ready to use.

PICKLED WINTER COLE PEARS

Pickling pears is a great way to use up the excess crop, and a little chilli helps to cleanse the palate. Serve alongside roasted meats, especially pork and smoked chicken.

1 cinnamon stick
6 cloves
1 tablespoon fresh ginger
600 ml white wine vinegar
700 g sugar
2 kg Winter Cole pears, peeled, cored and cut into wedges
dried chillies (optional)

Tightly tie cinnamon, cloves and ginger in a small piece of muslin. Bring the vinegar and sugar to the boil with the spices and leave to infuse. Add the pears to the vinegar and simmer for 10 minutes until they are tender but still firm. Remove the pickling spices. Pack into sterilised jars and if desired, add 1 dried chilli to each jar before sealing tightly using the hot-pack method (see page 13). Leave at least 2 weeks before using.

PICKLED RED CABBAGE

This recipe brings out the real flavours of red cabbage through the technique of salting to withdraw excess moisture. Ensure you use a plastic or other non-reactive metal (aluminium, tin and copper are reactive metals) vessel to drain the red cabbage, otherwise the cabbage will be tainted and lose colour.

- 1 medium-sized red cabbage
- 3 tablespoons salt
- 2 tablespoons sugar
- pickling spices such as peppercorns, juniper berries, or allspice berries
- 1¼ cups cider vinegar

Cut the cabbage into quarters and discard the tough central core. Shred the cabbage finely and place in a large non-reactive metal vessel or plastic container; add the salt and mix through the cabbage. Place a weighted plate on the cabbage and leave for 24 hours to draw out excess moisture. Rinse the cabbage thoroughly and drain on paper towels. Pack tightly in a large, warm, sterilised jar, layering with sugar and spices. Cover with vinegar and seal using the overflow method (see page 14). Leave for 2–3 weeks before serving with pork or cold cuts of meat.

RHUBARB & ORANGE CHUTNEY

The addition of oranges makes this chutney zesty and fresh. It's perfect with smoked chicken or poached pork.

5 g allspice
20 g mustard seeds
20 g black peppercorns
1 kg rhubarb
juice and zest of 2 oranges
500 g onions, diced
800 ml brown vinegar
900 g brown sugar
300 g seedless raisins

Tie the spices in a muslin bag and place all ingredients in a large saucepan. Bring to the boil and simmer gently for 1½ hours, uncovered, until the mixture is thick. Stir occasionally to prevent the mixture sticking to the bottom. Remove the muslin bag and spoon the chutney into hot sterilised jars. Seal using the hot-pack method (see page 13) and store in a cool place.

SPICED PEARS

Pears work well with poultry and game meats, and the sweet-and-sour effect of the vinegar and sugar ensures the fruit is not overly sweet.

3 kg firm cooking pears, peeled, cored and quartered
4 cups white wine
4 cups white wine vinegar
225 ml water
1 kg sugar
3 lemons
zest of 1 orange
8 cloves
150 g honey
2 tablespoons mustard powder

Combine all ingredients in a large, heavy-based saucepan, bring to the boil and simmer for 30–40 minutes until the mixture has thickened. Pour into hot, sterilised jars, seal using the hot-pack method (see page 13) and store in a cool, dark place.

PRESERVED LEMONS

I love these and always have large jars of them in my kitchen. Not only do they look good, but they provide a lemony hit when the lemon tree is not performing as expected. This is a traditional Middle Eastern form of preserving and is well worth the effort. Choose good lemons, as you will only be using the flesh.

For every lemon you're preserving, you'll need:

water
1 teaspoon salt, plus 1 extra teaspoon per lemon for the jar
juice of 1 lemon

Soak lemons in water for three days, changing the water each day. Make 4 incisions in each lemon by running the knife all the way down the lemon, but keep it intact at each end. Then gently place 1 teaspoon of salt inside. Repeat with the other lemons. Pack into sterilised jars, and make up a salt brine of equal proportions with the salt and 2 litres of boiling water (i.e. 20 lemons, 20 teaspoons salt, juice of 20 lemons). Pour this liquid over the lemons and seal tightly using the overflow method (see page 14). Store for at least 6 weeks, and use the skin only for adding to vinaigrettes, grilled fish and chicken.

SMOKED ONIONS

Try these tasty titbits when you need to add more flavour to your meals or salads. Smoke the onion in large pieces, but when serving slice them thinly. They'll keep for 7 days when covered in the refrigerator.

8 medium-sized white onions
1 tablespoon olive oil
1 teaspoon salt
½ teaspoon black pepper
wood chips, such as cherry or manuka, soaked in water for 1 hour

Peel the onions but leave the stalk intact as this will prevent them from falling apart. Cut into halves and then quarters, and brush with olive oil, salt and pepper. Using a smoker or two large roasting trays, put the onions on the rack and wood chips underneath. Cover with a lid and bring up to a hot smoking point, then turn the heat down as low as it will go and leave for 10–12 minutes. Store in an airtight container in the refrigerator or cover in grapeseed oil — the smoke flavour will infuse into the oil, so you can use the oil as a dressing.

BRANDIED ORANGES

Oranges steeped in brandy are a fantastic special dessert when time is limited. Serve them with whipped cream and small shots of the reserved brandy — your guests will never know just how easy it was.

500 g juicy, ripe oranges, well washed and scrubbed
400 g sugar
½ cinnamon stick
3 cloves
3 cups brandy
peel of 2 lemons

Pierce each orange 8 times and place them into a sterilised, wide-mouth jar. Add the sugar, cinnamon, cloves and lemon peel and pour the brandy on top until the oranges are submerged. Seal using the overflow method (see page 14), shake the jar until the sugar dissolves and leave in a cool, dark place. Shake the jar once a week for 6–8 weeks to ensure the sugar does not settle.

PARMESAN OIL

This is a great way to use up scraps of Parmesan, or the waxy end piece that you can't eat. Fill a small jar to halfway with grapeseed or rice bran oil (both have no flavour or greasiness) and as you finish your wedge of Parmesan cheese, place the scraps in the jar. After a couple of weeks, the oil will take on the Parmesan flavour and you can start to use it to dress salads or garnish fish. As long as the scraps remain submerged in the oil and it is kept in a cool, dark place, you can keep topping up the jar with oil and cheese scraps and using over and over again.

LEMON OIL

This needs to be kept either in a dark glass bottle or in the pantry, as light will affect the oil and lemon. Lemon oil goes well with all fish, seafood and lamb, and as a base for salad dressings. I use grapeseed oil or rice bran oil, as it has no greasiness or flavour to interfere with the taste of the lemon.

1 cup of grapeseed oil or rice bran oil
12 lemons, sliced

Heat the oil and lemons to 80°C in a large saucepan and keep it at that temperature for 30 minutes. Pour into a sterilised jar and top up with oil if needed. Seal with an airtight lid and keep in a cool, dark place.

LEMON CURD

Use unwaxed lemons in season to make this sweet, tangy spread, then enjoy it at the breakfast table or for dessert with poached pears or plums. The curd will keep for 3 months, but once opened must be used within 14 days.

450 g sugar
juice and grated zest of 4 lemons
110 g butter
4 eggs, beaten

Place the sugar, juice and zest, butter and eggs in a large bowl sitting on top of a saucepan of simmering water. Stir with a wooden spoon until thick and curd coats the back of the spoon. Pour into warm sterilised jars, seal using the hot-pack method (see page 13) and store in the refrigerator.

LEMON SYRUP

Refreshing and a great way to use up the excess of lemons. Try with limes or oranges instead, and mix with water to taste when drinking. This syrup can be used as a pouring syrup over fresh fruits or puddings.

1 kg caster sugar
600 ml water
900 ml lemon juice
zest of 2 lemons

Bring all ingredients to the boil and simmer for 3 minutes before pouring into warm sterilised bottles. Seal with an airtight lid and store in a cool, dark place.

CRANBERRY CHEESE

Cranberries are growing in popularity — we can now buy fresh cranberries in season, as well as cranberries in juice, sauces and dried form. Enjoy cranberry sauce with white meats such as turkey or chicken during the festive season.

3 kg cranberries
3 kg sugar
½ cup orange juice
zest of 1 orange
1 tablespoon lemon juice
1 teaspoon ground cinnamon

In a large saucepan, cover the cranberries with water, bring to the boil and simmer until soft. Pass through a mouli or sieve into a clean saucepan. Stir in the sugar and bring back to a simmer, stirring frequently. Add the remaining ingredients and simmer until the cranberry cheese is stiff and a wooden spoon leaves a trail through the mixture. Pour into small, hot sterilised jars and seal using the hot-pack method (see page 13).

CRANBERRY SYRUP

Enjoyable when added to any dessert, this tangy syrup is packed with colour and flavour. Sprinkle it with some chopped fresh mint when serving, to freshen it up further.

1 cup fresh or frozen cranberries
½ cup maple syrup
⅓ cup sugar
⅓ cup water
1 tablespoon lemon juice

Combine all ingredients except the lemon juice in a saucepan and bring to the boil for 10 minutes, skimming off any fruit scum that comes to the surface. Mix in lemon juice, pour into hot sterilised jars or bottles and seal with an airtight lid. Will last 14 days kept in the pantry or 3–4 weeks in the refrigerator.

BITTER ORANGE MARMALADE

Bitter Seville oranges are used in many ways, although they're not eaten raw — instead they are used for essential oils in perfumes and as herbal medicines. The bitterness counteracts the sugar in this marmalade, which can be spread liberally on buttered toast or used to finish off marinades or sauces.

1.5 kg Seville oranges
2 large lemons
14 cups water
3 kg sugar

Peel the oranges and lemons and slice the peel into strips. Juice the fruit, place the pips and pulp in a muslin bag and tie firmly. Simmer the juice, zest, water and muslin bag until the liquid is reduced by half. Leave to cool overnight. The next day remove the muslin bag, squeeze it out and bring the liquid back to the boil. Add the sugar and boil rapidly for approximately 10 minutes until it has reached setting point. To test for setting point, pour a little marmalade onto a cold plate, let it cool, then run your finger through it — there will be a trail left through the marmalade if it is ready. Pour into hot sterilised jars and seal using the hot-pack method (see page 13).

RHUBARB COMPOTE

To be enjoyed any time of the year in hot fruit pies or with slices of cake, rhubarb is one of the most overlooked ingredients in any garden. If the stalks are thick and woody, peel them first before cutting.

5 cups rhubarb, cut into small strips
½ cup sugar
½ teaspoon orange or lemon zest
¼ cup water
¼ cup currants
1 cup peeled and sliced apple

Place all the ingredients in a heavy-based saucepan, bring to the boil and simmer for 8 minutes, stirring often. Pour into hot sterilised jars and seal using the hot-pack method (see page 13). Store in a cool, dark place. Will last for 8–12 months and best eaten within 7 days once opened.

SAVOURING THE FLAVOURS

SAVOUR THOSE FLAVOURS year-round with these suggestions on how to use and serve some of your delicious home-made preserves.

SPICED BLUEBERRY & LEMON CURD FRIANDS

These friands are a nice alternative to heavy, starchy muffins and can be made with any fruits. They will keep for 4–5 days in an airtight container.

1 cup icing sugar
1 cup ground almonds
⅓ cup flour
4 egg whites
¼ cup melted butter
1 cup Spiced blueberries, drained (see page 67)
¼ cup Lemon curd (see page 113)

Sift half a cup of icing sugar, with the ground almonds and flour. Whisk the egg whites with remaining half cup of sugar until it stands in stiff peaks. Fold into the sifted ingredients with the melted butter. Spoon the mixture into small greased muffin tins, about two-thirds full and sprinkle with spiced blueberries. Bake at 180°C for approximately 12 minutes until cooked. Turn out onto a cake rack and allow to cool. Top with lemon curd and dust lightly with icing sugar.

CHILLED TOMATO SOUP WITH ROCKET OIL

Serve chilled on a balmy evening, this soup is the very essence of summer and simplicity. Rocket oil can be substituted for any other flavoured herb oil.

2 kg overripe tomatoes
1 teaspoon salt
½ teaspoon freshly ground black pepper
200 ml vegetable stock (either homemade or bought)
Rocket oil (see page 60), to garnish

Purée the chopped tomatoes and seasoning in a food processor until smooth. Slowly add the vegetable stock. Pour into a muslin bag and hang overnight in the refrigerator above a container to catch the clear juices that ooze out. In the morning you will be left with an intensely flavoured, clear tomato soup. Season to taste and serve in wide-mouthed wine glasses. Garnish with a good teaspoonful of rocket oil in each glass.

PICKLED WALNUT MAYONNAISE

Pickled walnuts have a most interesting taste and texture. The black colour of this walnut mayonnaise complements cold cuts of meat, game and pasta dishes. It is easy to prepare and will last for 5 days covered in the refrigerator.

3 free-range egg yolks
4 Pickled walnuts (see page 70)
pinch of salt
1 teaspoon lemon juice
1 teaspoon Dijon mustard
½ cup olive oil
1½ cups grapeseed or vegetable oil
2 tablespoons boiling water

In a food processor, or by hand, whisk the yolks, walnuts and salt with the lemon juice and mustard. Slowly pour in the oils, processing or whisking all the time, until thick and creamy. Add the boiling water and season to taste.

PRESERVED CHERRY TARTLET

This is an ideal way to use your preserved cherries in the middle of the winter — and the smell of freshly baked pastry wafting through the kitchen is especially enticing.

Sweet shortcrust pastry (either homemade, page 119, or readymade)
100 g butter
100 g caster sugar
2 eggs
3 tablespoons flour
130 g ground almonds
1 cup of preserved cherries (use either Cherries in cassis or Brandied cherries, both on page 59)

Line a 20-cm loose-bottomed flan tin with sweet shortcrust pastry and bake blind. Reserve some pastry for the lattice top. While the pastry cools, cream the butter and sugar until pale. Add the eggs and beat well, sift in the flour and ground almonds and mix to make a paste. Spread a quarter of the mixture into the cooked pastry case and then add cherries. Top with the remaining sugar/almond paste. Cover with a lattice of uncooked pastry and bake for 25 minutes until springy and firm. Dust with icing sugar and serve with freshly whipped cream.

SWEET SHORTCRUST PASTRY

200 g unsalted butter
100 g caster sugar
1 egg (optional)
300 g plain flour

Cream the butter and sugar until the mixture is a pale yellow. Add the egg, if using, then fold in the flour and mix to a firm paste. If you are not using the egg, add 2 tablespoons cold water. Form into a ball and flatten before covering with plastic wrap and resting in the refrigerator for at least 15 minutes. This will make it easy to roll out onto a floured surface before lining the pie or flan dish.

SMOKED ONION & LEMON RISOTTO WITH PARMESAN CHEESE

Hearty and warming, this risotto will really hit the spot. It can be made in advance and chilled for eating later. For variety and more flavour, experiment by adding either Wine-barrel-smoked mushrooms (see page 95), Pickled mushrooms (see page 91) or Mushrooms preserved in white wine & thyme (see page 90).

4 cups (or more) vegetable broth or stock
2 tablespoons butter
300 g Smoked onions (see page 110)
1 cup arborio rice
½ cup dry white wine
¼ cup freshly grated Parmesan cheese
2 tablespoons mascarpone cheese
1 tablespoon Preserved lemon zest (see page 110) or fresh lemon zest
4 tablespoons chopped parsley

Bring the broth or stock to a simmer in a saucepan. Reduce heat to low and keep warm. Melt the butter in a large saucepan over a medium heat. Add smoked onions and rice, then the wine and cook until almost all the liquid is absorbed, stirring frequently. Adding 1 cup of broth or stock at a time, cook until almost all broth is absorbed before adding more, stirring frequently, until rice is tender but still firm, about 20 minutes. Add the Parmesan, mascarpone, lemon zest and parsley. If dry, add more broth by quarter cupfuls as needed. Season with salt and pepper to taste.

MARINATED GLOBE ARTICHOKES & FETA QUICHE

This is a tasty alternative to a pie, and it's easy to make and never fails.

1 tablespoon olive oil
½ medium onion, diced
2 cloves garlic, minced
14 Marinated globe artichokes (see page 49)
4 eggs, beaten
2 tablespoons flour
¼ teaspoon oregano, fresh or dried
1 tablespoon fresh parsley, chopped
¼ cup milk
½ cup crumbled feta cheese
¼ cup shredded Parmesan or cheddar cheese

Heat the olive oil and sauté onion and garlic until soft. Drain and chop the artichokes, add and heat through. In a large bowl beat together the eggs, flour, oregano, parsley, and milk. Stir in the vegetable mixture and cheeses. Pour into a greased flan or quiche dish and bake at 350°C for 20–25 minutes or until set in the centre. Serve with a green salad or roasted red peppers.

CHILLI SALSA DROP SCONES

1 cup all-purpose flour
½ cup wholewheat flour
2 tablespoons wheat germ
2 teaspoons baking powder
½ teaspoon baking soda
½ teaspoon salt
6 tablespoons butter
about ¾ cup buttermilk
¼ cup Chilli salsa (see page 48)

Sift all dry ingredients in a bowl and rub in butter until it resembles coarse breadcrumbs. Make a well and stir in the buttermilk and Chilli Salsa to make a dough that can be spooned but still holds its shape. Drop dough by large spoonfuls onto a greased baking tray. Dust lightly with flour. Bake at 180°C for about 12 minutes, or until golden.

HONEY-SPICED APRICOT & COCONUT CRUMBLE

A quick and easy way to prepare any preserved fruits, but is particularly good with Honey-spiced apricots. Serve hot or cold with lashings of whipped cream, ice-cream or custard.

- 1 cup plain flour
- 1 cup rolled oats
- 1 cup brown sugar
- ½ teaspoon cinnamon
- 250g butter
- 1 cup thread coconut
- 4 cups Honey-spiced apricots (see page 66), or similar fruits

Mix flour, oats, brown sugar and cinnamon together and then rub in butter. Add the coconut. Place the apricots into a buttered or greased ovenproof dish and cover with crumble mixture. Bake for 20 minutes at 180°C until golden brown. Serve with custard or a fruit couli, or whipped cream or ice-cream.

ORANGE MARMALADE & WALNUT CAKE

This is a great 'anytime' cake with lots of flavour. It's a great way to use any flavoured fruit preserve — try substituting the orange marmalade with Cherry jam (see page 61) or Cranberry cheese (see page 114).

- 2/3 cup butter or margarine, softened
- 2/3 cup caster sugar
- 1 tablespoon lemon juice
- 1/2 cup Bitter orange marmalade (see page 115)
- 3 large eggs
- 1 1/2 cups all-purpose flour
- 1 teaspoon baking powder
- 1 teaspoon cinnamon
- 1/4 teaspoon ground cloves
- 1/4 teaspoon nutmeg
- 1/3 cup milk or buttermilk
- 1/2 cup walnuts, chopped

Cream the butter and sugar, then add the lemon juice and marmalade. Add the eggs and whisk well. Add sifted flour, baking powder and spices. Mix well and add the milk and nuts. Pour into a 22-cm cake tin lined with greased paper, bake at 170°C for approximately 50 minutes or until a skewer can be removed cleanly. Allow to cool for 10 minutes and then turn onto a cake rack. Serve with thickened cream and a sprinkle of cinnamon sugar.

PRESERVED LEMON OR LIME COUSCOUS

Light and nutritious, this goes well with all seafood and poultry dishes.

- 2 cups couscous
- 2 cups water
- 2 tablespoons olive oil
- salt & freshly ground black pepper
- 1 bunch spring onions, finely chopped
- 2 tablespoons finely chopped Preserved lemon (see page 110) or Dry-salted limes (see page 73)
- 4 tablespoons finely chopped fresh parsley

Mix couscous and water and allow to stand covered for 5 minutes. Add olive oil and fluff with a fork, season and add remaining ingredients. Will keep for 4–5 days covered in the refrigerator.

REFERENCE GUIDES

QUICK GUIDE TO SEASONAL PRODUCE

NB: This book does NOT contain recipes for ALL the fruits and vegetables in this table. It is intended as a guide only.

Produce	Peak season	Methods of preserving
Apples	Summer–autumn	Bottling, butter, cheese, chutney, drying, freezing, jam, jelly, pickling
Apricots	Summer	Bottling, cheese, chutney, drying, freezing, jam, pickling
Artichokes: globe Jerusalem	 Late spring–early summer Winter	 Bottling Pickling
Asparagus	Late spring–early summer	Bottling, freezing, pickling
Aubergines	Late summer–autumn	Bottling, pickling, salting, smoking, chutney
Avocados	Autumn	Freezing
Bananas	Summer	Chutney, pickling
Beans: broad French (green)	Spring- summer Summer–early autumn	Bottling, freezing Bottling, freezing
Beetroot	All year	Bottling, pickling, relish, drying
Blackberries	Autumn	Bottling, freezing, jam, jelly, using alcohol, using vinegar, cheese
Blackcurrants	Summer	Bottling, freezing, jam, jelly, using alcohol, using vinegar
Blueberries	Summer	Bottling, freezing, jam, jelly, pickling
Boysenberries	Summer	Bottling, freezing, jam, jelly
Broccoli	All year	Bottling, freezing, pickling
Brussels sprouts	Winter	Pickling, freezing
Cabbage	Spring, autumn & winter	Pickling, salting, bottling
Capsicums	Spring–autumn	Chutney, drying
Carrots	Winter–summer	Bottling, freezing, pickling
Cauliflower	Autumn–spring	Bottling, pickling, relish
Celeriac	Winter	Bottling, pickling
Celery	All year	Freezing, pickling, relish
Cherries	Spring–early summer	Bottling, freezing, jam, jelly, pickling, using alcohol, using vinegar

Produce	Peak season	Methods of preserving
Chillies	Spring–autumn	Bottling, pickling, relsih
Courgettes	Spring–autumn	Freezing, pickling, relish
Cranberries	Autumn–winter	Cheese, chutney, jam, jelly
Cucumbers	Spring–autumn	Pickling, relish
Fennel	All year	Bottling, freezing, pickling, relish
Figs	Summer–autumn	Bottling, chutney, drying, freezing, jam, pickling
Garlic	Late spring–summer	Bottling, pickling
Gooseberries	Summer–autumn	Bottling, chutney, freezing, jam, jelly,
Grapefruit	Winter	Jam, jelly
Grapes	Summer–autumn	Freezing, jelly, pickling
Kale	Winter	Freezing
Kiwifruit	Autumn–winter	Jam, drying
Leeks	Spring–autumn	Bottling, freezing, pickling
Lemons	Autumn–winter	Chutney, jelly, pickling, salting, using oil
Limes	Summer–winter	Chutney, jelly, pickling, salting
Loganberries	Summer	Bottling, jelly
Mandarins	Late autumn–early spring	Bottling
Melons	Summer–autumn	Jam, pickling, salting
Mushrooms	Autumn	Drying, freezing, pickling, smoking
Nectarines	Spring–summer	Bottling, jam, drying, chutney
Okra	Spring–autumn	Bottling, pickling
Olives	Autumn	Pickling, salting, curing
Onions	All year	Chutney, drying, pickling, relish, smoking
Oranges	All year	Chutney, drying, freezing, jam, jelly, using alcohol
Parsnips	Winter–spring	Freezing
Peaches	Summer	Bottling, drying, freezing, jam, pickling, chutney
Pears	Summer–winter	Bottling, drying, freezing, jam, pickling
Peas	Late spring–summer	Freezing
Persimmons	Autumn	Bottling, jam
Plums	Summer	Bottling, chutney, drying, pickling, jam, using alcohol

Produce	Peak season	Methods of preserving
Pomegranates	Autumn	Bottling, jam
Pumpkin	Summer–winter	Freezing, drying
Quinces	Autumn	Bottling, jam, jelly
Radishes	Spring–summer	Bottling, pickling
Raspberries	Summer	Freezing, jam, using alcohol, vinegar
Redcurrants	Summer	Freezing, jam, jelly, using alcohol
Rhubarb	Spring, autumn & winter	Freezing, jam, jelly, bottling, chutney
Shallots	Spring–autumn	Chutney, drying, pickling
Silverbeet	Summer-winter	Freezing
Spinach	All year	Freezing
Spring onions	Spring–autumn	Bottling
Squash	Spring–autumn	Bottling, freezing
Strawberries	Spring–summer	Freezing, jam, using alcohol, vinegar
Sweetcorn	Summer	Freezing, relish
Tamarillos	Autumn–winter	Bottling, chutney
Tomatoes	Summer–early autumn	Bottling, chutney, drying, freezing, pickling, smoking
Walnuts/green	Summer	Bottling, pickling
Watermelon	Summer	Bottling, salting

TEMPERATURE CONVERSION

Celsius	Fahrenheit	Gas mark
120°	250°	1/2
135°	275°	1
149°	300°	2
162°	325°	3
176°	350°	4
190°	375°	5

Celsius	Fahrenheit	Gas mark
204°	400°	6
218°	425°	7
232°	450°	8
246°	475°	9
260°	500°	9

INDEX